NAS Monograph Series No. 5

Series editor Gerald Grainge

Nautical Archaeology Society

The Gresham Ship Project

A 16th-Century Merchantman Wrecked in the Princes Channel, Thames Estuary

Volume II: Contents and Context

Edited by

Gustav Milne and Dean Sully

with contributions by
Mark Beattie-Edwards, Lynn Biggs, Thomas Birch,
Michael F. Charlton, Kelly Domoney, Clare Hunt, Phil Magrath,
Marcos Martinón-Torres and Zofia Stos-Gale

BAR British Series 606
2014

Published in 2016 by
BAR Publishing, Oxford

BAR British Series 606

Nautical Archaeology Society Monograph Series (NAS) 5

The Gresham Ship Project

ISBN 978 1 4073 1211 8

BAR Publishing is the trading name of British Archaeological Reports (Oxford) Ltd. British Archaeological Reports was first incorporated in 1974 to publish the BAR Series, International and British. In 1992 Hadrian Books Ltd became part of the BAR group. This volume was originally published by Archaeopress in conjunction with British Archaeological Reports (Oxford) Ltd / Hadrian Books Ltd, the Series principal publisher, in 2014. This present volume is published by BAR Publishing, 2016.

Printed in England

BAR
PUBLISHING

BAR titles are available from:

BAR Publishing
122 Banbury Rd, Oxford, OX2 7BP, UK
EMAIL info@barpublishing.com
PHONE +44 (0)1865 310431
FAX +44 (0)1865 316916
www.barpublishing.com

Table of Contents

Acknowledgements

The Gresham Ship Project team warmly acknowledges the help of many institutions and individuals (regrettably there is not space here to mention them all). First and foremost must be the Port of London Authority who supported the project from the outset, through the good offices of Nicola Clay, David Cartlidge, Alistair Gale and especially Gill Andrews, who watched over the deliberations of our Steering Group with her customary efficiency and sound advice. The team of maritime archaeologists from Wessex Archaeology, directed by Tony Firth, recorded and recovered the wreck and its contents in far from ideal conditions in the Thames Estuary. Further welcome support was provided by Geoff Pavitt, Barbara Anderson and Valerie Shrimplin from Gresham College, from the Mercers' Company, from the English Heritage team at Fort Cumberland (Hampshire), from Phil Magrath and the Royal Armouries team based in the neighbouring Fort Nelson and from Alison Kentuck, the Receiver of Wreck. Staff from the Museum of London included Roy Stephenson, Hazel Forsyth and Tom Wareham, while Ken Crowe and Clare Hunt from the Southend Museums Service also willingly assisted.

The project would not have progressed without the facilities and generous spirit of the staff at the Universities concerned; especial mention must be made of Jens Auer (who first dived on the wreck over a decade ago and has championed its study ever since) and Professor Thjis Maarvefeld of the University of Southern Denmark, and of Professor Stephen Shennan, Professor Clive Orton, Barbara Brown, Louisa Goldsmith and Thom Rynsaard at University College London. This monograph benefitted from the editorial work of Sonia Skaraas and the illustrations of Jenny Stripe and Chris Unwin, while the authors of the works listed in the extensive bibliography contributed to the research, as did, knowingly or unknowingly, Robert Baldwin, Alan Bryson, Matthew Davies, Ian Friel, and Anthony Payne, our anonymous referees and particularly Gerald Grainge.

But this monograph owes a significant debt to the professionalism and dedication of the remarkably talented students who contributed so much to the research, the results of their labours are presented in Chapters 3 and 4.

For their contributions to the research reported in Chapter 3, Dean Sully and Kelly Domoney warmly acknowledge the support of English Heritage, through Jaquie Watson, Angela Karsten and Karla Graham, as well as from Jackie Keily, Hazel Forsyth, Geoff Egan and John Shepherd from the Museum of London. Their colleagues at UCL gave much needed assistance, especially Sandra Bond, James Hales and George Davies, as well as the army of graduate students who participated in the Gresham Ship Project as a part of the MSc Conservation for Archaeology and Museums programme. These include Natsumi Henzan, Sylvia Haliman, Lyda Muros, Cymbeline Storey, Luisa Duarte, Sarah Nettleton, Robyn Haynie, Rosemary Jeffreys, Alex Solomon, Ashley Lingle, Danielita Pedroza de Boos, Julie Chang, Amanda Watts, Morgan Nau, Rebecca Pont, Maria Vida Naverro and Rafael Lopez Bravo, while special thanks were earned by Hazel Gardiner, Libby McCormick, Katy Smith, Julia Tubman, Graeme McArthur, Nicole Daub, Kelly Hanson, Lisa Barter and Jill Saunders.

The study of the Gresham assemblage by Marcos Martinón-Torres and his team (reported in Chapter 4) would not have been possible without the help and assistance of many individuals, as well as generous support by the Port of London Authority and the project's partners. Many thanks are due to Stavriani Orfanou and Ioannis Papadias for their help in documenting and sampling the Gresham iron bars, as well as to English Heritage at Fort Cumberland for their support and assistance. In particular, we are grateful to Kelly Domoney for her time and expertise in the X-radiographic examination of some of the iron bar fragments.

The chemical results obtained from analysis at the Wolfson Laboratories of the UCL Institute of Archaeology benefitted immensely from the steadfast expertise, experience and technical assistance of Kevin Reeves, who also carried out the EPMA-WDS analyses of lead and tin samples. Simon Groom provided guidance in sample preparation. LIA and NAA analyses were supervised by Ernst Pernicka at the Curt-Engelhorn-Zentrum Archäometrie in Mannheim. We would also like to thank David Crossley, Gustav Milne, Thilo Rehren and Lynn Willies for their time in conversing over the results and for their useful suggestions; and Philippe Dillmann for providing SI reference data.

Finally Operation Gresham (the transfer of the Gresham Ship hull sections from Horsea Island Lake), organized by the infectiously enthusiastic Nautical Archaeological Society, would not have been possible with the help of both sponsors and volunteers (their names are fully listed by Mark Beattie-Edwards in Chapter 6). The NAS gratefully acknowledges the support of The Port of London Authority, the Mercers Company, University College London, University of Southern Denmark, the Stoney Cove Dive Centre, QinetiQ, the Royal Navy and the Field Squadron (Air Support) of 39 Engineer Regiment (based at Waterbeach Barracks in Cambridge), without whom the last voyage of the Gresham Ship could not have been launched.

Abstract

This is the second volume of a two-monograph set that describes and discusses the Gresham Ship, an Elizabethan armed merchantman built *c.* 1574 that had sunk in the Thames Estuary some 20 to 30 years later. The first volume describes and discusses the recording and recovery of the hull remains and associated objects, such as guns. This initial work was undertaken by a team from Wessex Archaeology in 2003–4, working with the Port of London Authority. The first volume of the monograph presents the detailed research that subsequently took place in the University of Southern Denmark (2007–2013), analysing, modelling and interpreting the vessel represented by the five substantial hull fragments recovered, comparing the results with broadly contemporary vessels found on other wreck sites.

This second volume presents the studies conducted at University College London. Chapter 1 provides a summary of the Gresham Ship Project, a university-based project co-ordinated by University College London, and supported by the Port of London Authority from 2007 to 2012. In Chapter 2 Gustav Milne considers the topography and workings Tudor Port of London, the harbour which may well have served as the home port for the Gresham Ship.

The next two chapters deal with the finds from the wreck site. In Chapter 3, Dean Sully and Kelly Domoney look at the assemblage from a conservation perspective, describing innovative approaches to the work on the concretions recovered from the bed of the Thames and on a range of small finds of metal, wood and leather.

The next chapter considers the cargo from the wreck site, with a detailed description of the metallurgical analysis of the lead and tin ingots, as well as the bar iron. A provenance for the metalwork is suggested and its significance assessed by Thomas Birch and Professor Marcos Martinón-Torres.

Chapter 5 is an essay on vessels like the Gresham Ship, armed merchantmen of the late 16th century; such craft were used as traders, for distant exploration of New Worlds, for privateering or to support the Queen's navy in the war against Spain.

This is followed in Chapter 6 by a report from Mark Beattie Edwards on how the Nautical Archaeology Society organized the complex transfer of the hull remains from a lake in Hampshire to a flooded quarry in Leicestershire. The remains of the Gresham Ship now serve as a unique underwater museum exhibit at the National Dive Centre in Stoney Cove, as well as an underwater classroom in which the next generation of nautical archaeologists can learn their craft.

This is followed by a consolidated bibliography for this second volume. There is also an Appendix which includes a catalogue of the recovered artefacts, listing which ones are now at Stoney Cove and which ones are in the care of the Southend Museum Service, awaiting display in a purpose-built gallery in a brand new museum building, now being planned.

Chapter 1: The Gresham Ship Project 2007–2012

The Background

In May 2012 an armed Elizabethan merchantman began its final voyage. Some 400 years after it sank in the Thames estuary, its remains were carefully stacked onto two low loaders and were driven overland to their final resting place. The ship is now part of a unique underwater museum display in the National Dive Centre at Stoney Cove, where it also serves as the main focus for training the next generation of nautical archaeologists (see Beattie-Edwards, this volume, Chapter 6). But how did this remarkable exemplar of London's maritime history, a 16th-century sea-going ship from the golden age of Drake, Raleigh and Frobisher, end up in a disused quarry in Leicestershire?

This particular tale begins, not at the beginning, but in 1846, when the first recorded underwater excavation took place in the Thames Estuary. This pioneering project utilised hard-hat dive technology, newly developed at ports such as Whitstable by Charles and John Deane. The brothers were born in Deptford, schooled in Greenwich and worked in Barnard's Shipyard on the Thames. In 1820 Charles responded to a serious fire in a stable block by donning a metal knight's helmet and taking air from a hose and valiantly rescued all the horses from the smoke-filled building. His attempts to develop his impromptu fire-fighting equipment brought little financial reward and so the brothers reworked the idea as a surface-supplied underwater diving helmet. This was a notable improvement on the cumbersome diving bells previously used to investigate wrecks or other obstructions underwater. From this their own diving industry was developed in Whitstable and then in Gosport. In 1834–5 they were working for the Admiralty on the wreck of HMS *Royal George*, a 108-gun warship that had sunk in the Solent in 1782 in 60 ft of water, from which 28 guns were recovered. In July 1836 a group of local fishermen then asked them to investigate an obstruction on the seabed that was snagging their fishing lines. John Deane obliged and duly discovered ancient ship timbers protruding from the seabed and a 16th-century bronze gun: it was the long-lost site of the *Mary Rose*, Henry VIII's flagship, tragically sunk in 1545.

Salvage work on this new site continued intermittently until 1840 (sometimes making use of explosives) with many finds recovered. The guns were sent to the Office of Ordnance at Woolwich in 1840. As for the anchor, this was sent initially to Beale's Wharf in Southwark, from where it was to be sent to Sheffield for 'recycling'. An employee of the wharf, one A. Hillhouse was struck by the historic nature of the artefact and wrote to the Duke of Wellington hoping he might 'rescue so interesting a relic from its probable fate' (Marsden, 2003, 29). Since the letter still survives in his archives, it clearly caught the attention of the venerable old soldier, but it was not sentiment that interested him. By 1846 he was Lord Warden of the Cinque Ports (among many other titles), an office that still held responsibilities for the clearance of wrecks from sea lanes. Any material recovered from these sites – like anchor from the *Mary Rose* – could be salvaged and sold to pay for the work. And that may have been on his mind when news of another wreck containing a valuable cargo of metalwork was brought to his attention.

This new site was in the Thames Estuary, not far from Whitstable where the Deanes first developed their underwater salvage technology. As it transpired, the vessel proved to be another Tudor shipwreck, but the aim of the proposed underwater investigation was simply to salvage as much of its valuable metal cargo as possible. A contemporary comment noted: 'It had been known long since there was a wreck on the Girdler Sand (off Herne Bay), but no one took any notice of it, not knowing what wreck it was, until this spring, when divers went down and examined and recovered some iron guns, of very ancient date, also some of those curious ingots and some iron, lead in pigs, and red lead in cast iron casks, covered with wood'. The subsequent salvage work 'recovered about 2700 of the ingots, and more iron, pig lead, red lead, together with some stone shot.' (British Archaeological Association, 1847, 361)

The historic significance of the discovery was not completely overlooked, however; for at a meeting of the British Archaeological Association in December 1846, a handful of items recovered from the wreck were discussed. They included: 'an ingot of tin stamped with the royal mark, a rose surmounted by a crown; a knife with the maker's mark, a double fleur-de-lis, on the blade; a round-toed leather shoe and a dagged or pounced silk doublet', observed to be 'of a kind in fashion in the time of Elizabeth' (British Archaeological Association, 1847, 361–2). Provisional assessment of the material recovered therefore suggested an Elizabethan date for the ship. However, once sufficient metalwork had been recovered and recycled, interest in the site waned and the precise location lost.

2003–4 Rediscovery and Recovery

When the Port of London Authority (PLA) began clearance work in the Princes Channel in 2003, to facilitate the passage of modern shipping in the Thames, the remains of the 16th-century ship were dramatically rediscovered. The site lay some 11 km off Whitstable on the Kent coast and 22 km from Southend (Essex). A second underwater excavation project then began, but this time, 150 years later, with rather different aims, technologies and no explosives (Auer and Frith, 2007; see also Volume I, Chapter 3.2, page 20). Maritime archaeologists from Wessex Archaeology led by Jens Auer (now Assistant Professor of Archaeology at the University of Southern Denmark) were called in to evaluate and record the hull remains and associated artefacts. Iron bars, ingots of lead and tin, pottery and leather were also recovered, together with four guns, one of which had the 'TG' insignia on the barrel as well as the moulded emblem of a grasshopper. This is the mark of Sir Thomas Gresham (1519–1579), the famous London financier, merchant and gun-founder. The gun was presumably made at Gresham's foundry in Mayfield, which was in operation from 1567. Further confirmation of an Elizabethan date for the ship was provided by Nigel Nayling's assessment of twelve dendrochronological samples from the hull, suggesting that the timbers were felled in England (probably in Essex or East Anglia) in or shortly after 1574 (Volume I, Chapter 3.5, pages 43–5).

Once the fieldwork in the Thames estuary was complete, the five substantial sections of the carvel-built hull were raised from the sea bed. They represented a *c.* 5-m section of the bow and a *c.* 15-m length of part of the port side of the vessel from the keel to the level of the orlop deck and included elements of two gunports. The remains were initially transported to Denton Wharf (Gravesend) where further recording took place and then to a brackish water lake owned by the Ministry of Defence in Horsea, Hampshire.

The Gresham Ship Project Established

Deliberations then began as to how such an important, but unexpected find should be dealt with. Appreciating the unique significance of the discovery for nautical archaeology, as well as for London and the long history of its port, the PLA generously agreed to support an innovative five-year research programme coordinated by University College London. Studies of the hull and the ship's contents were to be integrated into graduate teaching and dissertation programmes so that graduate students would be provided with real, relevant hands-on experience to complement and extend their lecture-based studies. UCL was selected as the host, given the Institute of Archaeology's long association with maritime archaeology, with the Nautical Archaeology Society and with the study of London.

Work commenced in 2007–8 and a steering group was set up to oversee the programme, expertly chaired by Professor Clive Orton, with Gustav Milne and Dean Sully as project managers for UCL and Jens Auer from the University of Southern Denmark. There was representation from the PLA (Gill Andrews), the Receiver of Wreck (Alison Kentuck), the Nautical Archaeology Society (Mark Beattie-Edwards), the Museum of London (Hazel Forsyth and Roy Stephenson), the Royal Armouries (Phil Magrath) and Gresham College (Geoff Pavitt). The research was conducted partly in the UK, co-ordinated through the Institute of Archaeology at University College London, and partly in the University of Southern Denmark.

The Research Monographs

Two research volumes were compiled, reflecting the studies conducted at the two universities.

Monograph I: Excavation and Hull Studies

The first monograph describes the initial discovery and recording of the vessel remains in 2003–4. Research on the detailed records of the five substantial sections of the vessel itself were then conducted at the University of Southern Denmark, using computers and 1:10 scale models through a series of graduate projects supervised by Jens Auer. These studies show that the vessel was of about 180 tons burthen or tonnage of *c.* 250. It was 24.7 m in length at deck level with a keel 19 m long and an internal width (beam) of *c.* 7 m. The hold was some 3.2 m deep and the carvel-built hull was strongly framed, furnished with integral gunports. It seems clear that the vessel had been reworked to increase the beam and capacity of the hull (or as a major repair or renovation), a process later referred to as 'furring': this is the first time such a procedure has been identified archaeologically and is considered in detail. Four guns were recovered, two cast iron falcons, a cast iron minion and a wrought iron perrier; these are described, discussed and illustrated. They are currently in the care of the Royal Armouries at Fort Nelson, undergoing a long desalination treatment. The monograph concludes with an assessment of the Gresham Ship's construction in comparison with other contemporary vessels to establish how 'typical' or how 'English' it seems to be. The final chapter discusses the results and issues such as the particular circumstances of how this project developed and the role of universities in such research. Although the largest vessels of this period were twice this size, the Gresham Ship would not have been considered as a small ship; it was, for example, broadly the same size as the *Golden Hind* which successfully circumnavigated the globe in 1577–80. It is known that at least fifteen vessels of a similar size were built in London and the East coast region from 1574–7 and thus our ship could be seen as a typical, hard-working armed merchantman of this era.

Monograph II: Contents and Context

The majority of the finds assemblage was transferred to UCL in April 2009 from the English Heritage facility at Fort Cumberland, Hampshire. The work then focused on investigative conservation revealing the contents of 22 marine concretions (extensive deposits that have developed around artefacts on the sea bed), followed by the stabilisation of the recovered artefacts. This programme was supervised by Dean Sully, with notable contributions by Kelly Domoney, Libby McCormick and other UCL graduates working in the Conservation Labs.

One of the largest concretions contained a powder chamber, representing an additional gun, while painstaking work on another concretion revealed part of a wooden gun carriage. Other significant (but rather smaller) finds that have now been cleaned, processed and conserved include ceramics, a pewter salt holder, a silver spoon, a copper alloy dish. Organic materials such as leather have also recorded, including part of a seaman's boot, perhaps hurriedly kicked off as its owner struggled

Studies in the archaeology and history of the 16th-century port	
CONFERENCE May 16th 2009	
The Tudor Port of London: an archaeological investigation	
Museum of London in Docklands	
Chaired by Professor Clive Orton and Dean Sully	
Gustav Milne:	*Topography and Archaeology of the Tudor Port*
Geoff Pavitt:	*Thomas Gresham and the Royal Exchange*
Roy Stephenson:	*London's Tudor Naval Victualling Yard*
Jens Auer:	*Ships and Shipping in Elizabethan London*
Damian Goodburn:	*Boats and Barges in Elizabethan London*
Hazel Forsyth:	*Sugar and Spice and all things Nice*
Geoff Egan:	*Trade in the Tudor Port*
Jacqui Pearce:	*Imports and Exports*
Phil Magrath:	*London and the Gun Trade*
Frank Meddens:	*London's Pirate Community*
LECTURE SERIES 2008–9	
The lecture series on aspects of the Tudor Port included the following presentations:	
Professor G. Marnef:	*Thomas Gresham in Antwerp*
Professor T. Connell:	*Thomas Gresham and the New Learning*
Robert Baldwin:	*Queen Elizabeth I and the Politics of Prosperity*
Alan Bryson:	*Legal Quays: Sir William Paulet, first Marquis of Winchester*
Anthony Payne:	*Richard Hakluyt: London's role in Navigation and History*
Rear Admiral John Lippiett:	*The Mary Rose*
Professor T Connell:	*The Livery Companies in Tudor London*
Matthew Davies:	*Merchants and Heroes in the time of Stow*
Ian Friel:	*Elizabethan Merchant Ships and Shipping*

Table 1.1 *As part of a research and outreach programme designed to complement the Gresham Ship Project's work, Gresham College supported and hosted a conference and a lecture series.*

to leave the sinking ship. All this work is described and discussed in Chapter 3 of this volume. In Chapter 4 the focus is on the cargo of metals, a study supervised by Professor Marcos Martinón-Torres. Two types of bar iron were recovered, one folded once, the other folded four times, which would form a bar some 6 m in total length. The manganese content of some of these bars suggests they may have originated in the Rhineland. This particular type of bar iron, which is folded so as to be transportable by packhorse, is sometimes referred to as 'Voyage Iron' or 'Slave Iron', since it was used in the 17th century as currency exchanged for slaves from West African coast. The stamped lead ingots were boat-shaped and measured 600 x 220 x 120 mm, each weighing *c.* 56 kg, while the tin ingots were 520 x 18 mm. The isotopic signature suggests the tin is most likely of British origin, as is the best provisional match for the lead ingots.

Public Outreach

With generous support from Gresham College (the London institution founded as a bequest in the will of Sir Thomas Gresham), a series of public lectures on aspects of the Tudor Port of London was held in 2008–9, together with a full-day conference in the Museum of London in Dockland in May 2009 (Table 1.1). This was supported by a major display put together by a team of UCL's Museum Studies graduates, under the direction of Theano Moussouri and Ian Carroll, but led by Victoria Kinahan, with additional funding from the Mercers' Company. Artefacts from the ship also featured in a 2012–3 display celebrating the 75th anniversary of the founding of the Institute of Archaeology and its long-standing role in the promotion of the study of nautical archaeology.

The finds from the ship are now being transferred to the care of the Southend Museums Service. They will ultimately be displayed in a purpose-built gallery with a theme focussed on wrecks from the Thames, appropriately in a brand new museum building overlooking the estuary from which the Gresham Ship was recovered. Although that lies some years in the future, a temporary exhibition that featured some of the Gresham artefacts has already been presented in the Central Museum at Southend from June to October 2012.

In May 2012 the five sections of the hull were individually craned out of Horsea Lake and transported overland from Hampshire to Leicestershire to become a major exhibit in the National Dive Centre at Stoney Cove. This complex operation was directed by Mark Beattie-Edwards for Nautical Archaeology Society and by Martin Woodward for the NDC. This major move ensures that the remains will be in a secure, but accessible location, forming part of a major new heritage attraction – one of the UK's first underwater museums – for which underwater diver trails are being developed. The Gresham Ship is thus now on public display, albeit 6 m underwater, where the unique Elizabethan exhibit will be enjoyed by the some 30,000 divers of all ages who use the centre every year. The reconfigured wreck site is also used by the Nautical Archaeological Society to inspire and train the next generation of maritime archaeologists (see Chapter 6).

Chapter 2: The Port of Elizabethan London

by Gustav Milne with a contribution from Phil Magrath

Home port

The Gresham Ship was apparently built in an east coast yard, seems to have had a long and hardworking life and was eventually lost in the Thames Estuary. Since London was the largest port in England at the time (Broodbank, 1921, I, 45–57; Fisher, 1990), she must have visited the city often and indeed it is likely that the town was her 'home port' for at least part of her operative life (Fig. 2.1). The port's population increased from 50,000 to 200,000 in the 16th century, an astonishing increase, representing a burgeoning market. London's cosmopolitan nature is shown by a 1572 census, listing 3190 Dutch, as well as sizeable French and Burgundian communities; there were even 32 Scots (Weinstein, 1994, 3). The summary study of the town and the Thames presented in this introductory chapter tries to provide something of the topographical, mercantile, social and political background of the port. This forms part of the terrestrial context in which robust merchantmen such as the Gresham Ship operated. They were more than the mere workhorses of an increasingly prosperous and ever widening Tudor world; they were the engines of that expansion.

'Ryver of Thamys'

The Elizabethan antiquarian John Stow was a contemporary of Sir Thomas Gresham and, like him, was very familiar with London's harbour. In his great *Survey of London* finally published in 1598 he provides a valuable contemporary assessment of the port and the central importance of its river, which he compared most favourably to the other major waterways in England, writing: 'This realm hath only three principal rivers whereon a royal city might well be situated: the Trent, the Severn & the Thames.'

He praised:

> the straight course of the Thames, that its length reached furthest into the belly of the land, is most navigable up and down stream and that London standing almost in the middle of that course, is more commodiously served with provision of necessaries than any town standing on the other two rivers could be, and doth also more easily communicate to the rest of the realm the commodities of her own intercourse and traffic (Wheatley, 1956, 485).

Fig. 2.1: Tudor Port: a reconstruction of the busy 16th-century harbour by Chris Unwin, showing the relative sizes of the ships, boats and barges that jostled for space on the crowded river

He noted that the river was tidal for 'more than sixty miles in length, to the great commodity of travellers, by which all kinds of merchandise be easily conveyed to London, the principal storehouse and staple of all within the realm' (Wheatley, 1956, 13). Stow also understood impact of the port on its immediate maritime hinterland, stating that the city 'maintaineth in flourishing estate the counties of Norfolk, Suffolk, Essex, Kent and Sussex: they stand not so much by the benefit of their own soil as by the neighbourhood and nearness which they have to London' (Wheatley, 1956, 496–7).

As for its contacts with the continent, he goes on to say that the Thames 'openeth indifferently upon France and Flanders our mightiest neighbours, to whose doings we ought to have a bent eye and special regard: this City is not only near enough for intelligence of the affairs of those princes but also sufficiently removed from the fear of any sudden dangers that may be offered by them'. Summing up the situation of London on the great river, he made the classic comment that it was in the best possible location: 'for if it were removed more to the west, it would lose the benefit of the ebbing and flowing: if it were seated more to the east it should be nearer the danger of the enemy' (Wheatley, 1956, 485–6).

Working waterway

The many merchantmen working their way towards London in the 16th century would have the assistance of descriptive rutters (Waters, 1967) – many of which were now in print – as well as a new generation of recently-surveyed charts (Waters, 1958, 127–75; e.g. Pls 36, 41, 44, 50, 68). But the increased volume and size of shipping on the river that reflected the 16th-century economic expansion nevertheless required greater supervision and organization. Pilots provided by the Fellowship of Lodemen of the Cinque Ports, operating from Dover, guided merchantmen approaching London from the Channel coast. But from 1514 responsibility for pilotage in the often treacherous Thames estuary itself passed onto the Brotherhood of 'the Most Glorious and Undivided Trinity and of St Clement', an institution with religious origins then based in Deptford. In the aftermath of the Reformation it was subsequently reconfigured on more secular lines as a corporation known as Trinity House, committed to the 'science or art of mariners' (Harris, 1969). By the mid-16th century they had become involved in many river-related issues such as the provision and maintenance of beacons and buoys marking navigational hazards and also advised on the ordnance to be used for self-defence in merchant shipping. The corporation additionally concerned itself with matters relating to the supply of ballast (an issue with which it claimed to have been involved since the early 16th century (Harris, 1983, xi) and the conservation of the river. By the early 17th century their officials were recording the alarming rate at which the river seemed to be silting up; the Pool of London had been up to 13 ft deep (4 m), but was 8 ft (2.4 m) in *c.* 1620; at Limehouse it had been 16 ft (4.9 m), but was now 12 ft (3.7 m); at Blackwall it had been 19 ft (5.8 m), but was now 13 ft (4 m) (Harris, 1983, no 414, 121). Trinity House also had a role in the authorization of Thames watermen (Humpherus, 1859) and an Act of 1555 saw 'eight of the most wise, discreet and best sort of watermen' elected to supervise their peers to help prevent accidents 'caused by the ignorance or incompetence'. Licences were issued, ensuring that watermen were 'well able and sufficient' for their task, the size of the wherries was standardized and fees for journeys between Greenwich and Windsor were fixed (Broodbank, 1921, II, 379–80).

As for the larger merchantmen en route for London, once the Thames estuary was reached, the statutory pilot was taken on board for the statutory fee. Seamarks were provided by the churches whose towers and spires punctuated the skyline from the Reculver to St Paul's. For the first twenty miles or so both banks of the Thames were lined by inhospitable marshes and mudflats. But at Gravesend the river narrows and, since it was possible to make the first safe landing here, this is where London's outer defences were placed to protect it from invasion threats such as those of the 1540s, 1580s and 1590s. Five fortified gun batteries were built here, one pair on opposite banks at East Tilbury and Higham, another pair at West Tilbury and Gravesend, with a fifth at Milton (Smith, V., 1985; Moore, 2000).

The further west our vessel progressed, the more shipyards would be seen (Banbury, 1971). The Thames played a key role in the development of the 'Navy Royal', the pressing need for which developed with the century. Henry VIII's quarrel with the Pope had isolated Protestant England from powerful Catholic states across the narrow Channel, leaving the port and country vulnerable to sea-borne attack and the real possibility of invasion. The combined forces of France and Spain formed the principal threat from 1538 onwards, culminating in an attempted invasion in 1545. The threat was repulsed, although the Isle of Wight and Brighton suffered, but it led to an expansion of the fleet, with Deptford growing to become the Navy's largest yard by 1550, with Woolwich also much enlarged in 1546 (Moorhouse, 2005, 109). It was in these yards that new types of robust multi-masted, multi-decked warships were built. The *Mary Rose* is an example of the new design, in effect a merchant carrack design adapted for military duties; she was fitted out here on the Thames in 1512.

The sites of both these naval yards have benefitted from archaeological investigation. Evidence of 16th-century slips have been recorded at Woolwich (Courtney, 1974) while the foundations of a Tudor storehouse range at Deptford were recently revealed on the Convoy's Wharf site by a team from Museum of London Archaeology. These buildings must have been seen by Sir Francis Drake when he brought the *Golden Hind* to Deptford after his circumnavigation of globe (1577–1580). It was in Deptford that the famous Tudor shipwrights Matthew Baker and Peter Pett were working on such vessels as the

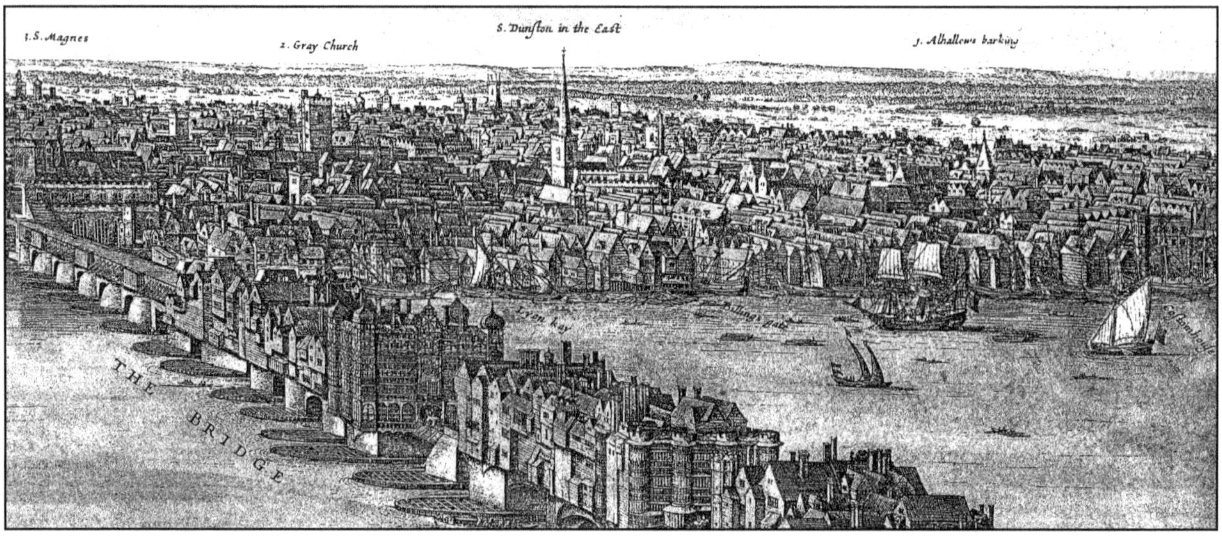

Fig. 2.2: *Elizabethan port of London: the area known as the 'legal quays' is shown, stretching from London Bridge (left) to the turreted brick-built Custom House (far right) as depicted by W. Hollar in 1647 (© London Metropolitan Archives, City of London). This was the busiest section of the harbour.*

Dreadnought, the *Handmaid*, the *Swiftsure* and the *Achales* (Oppenheim 1896, 120), all contemporaries of the Gresham Ship.

The last leg of the journey made by a merchantman en route to London passed through settlements of mariners, fishermen, boat builders, sail makers and rope makers that occupied both banks of the river immediately east of the City (McDonnell, 1978). Beyond that mix of modest dwellings and 'large and strong houses' built by shipwrights (Morris and Cozens, 2009, 2) was the great walled city itself. This was dominated by great span of London Bridge (Watson *et al.*, 2001), the towering bulk of minster church of St Paul's and the spires of 100 parish churches, as shown on contemporary panoramas (e.g. Colvin and Foister, 1996). One of the few elements of that vista surviving today is the Tower of London. This lies behind Tower Wharf, a river wall which still boasts its medieval ashlar facing at foreshore level, as described in a late medieval building contracts (Colvin, 1963, 726–7; Salzman, 1952, 468–70).

Just to the north-east of the Tower had been the precinct of the Cistercian abbey of St Mary Graces. Following the infamous Act of Supremacy in 1534, it was rudely transformed in the mid-16th century from a monastic house into the first large-scale supply depot and victualling yard for the new navy. A substantial part of the ground plan of the yard was recorded in the 1980s (Grainger and Hawkins, 1988) and it subsequently proved possible to identify slaughterhouses, salt houses, pickling sheds, bakeries, coopers' workshops, storehouses, offices and the dwellings of the work force (Grainger and Phillpotts, 2010).

London's Legal Quays: a port reformed

When the new Queen ascended to the throne in 1558, England was suffering from serious religious divisions and deep social, economic and financial problems. She addressed all these inter-related issues head on: for example, her very first parliament saw the Protestant Church of England established by law (Bindoff, 1950, 182–94) and measures were taken to introduce wide ranging changes to the tax collecting system (Dietz, 1972). The following year work began on reforming the coinage itself, an operation that successfully stabilized the currency and prices, while netting the state a profit of £14,000 (Bindoff, 1950, 190).

In London the reforms to the custom system in 1559 impacted positively on the Queen's personal finances and dramatically changed the topography of the port itself, not least with the building of a new red-brick Custom House. The reforms were instigated by the Lord Treasurer William Paulet (1474–1572), the Marquis of Winchester. He was remarkable not just for living to a very ripe old age, but for his ability to serve monarchs of all persuasions, working for Henry VIII, Edward VI, Mary Tudor and also Elizabeth herself. Such a track record bears testament to his forceful personality and an irreplaceable effectiveness in matters monetary and mercantile (Bryson, 2008).

From 1559 onwards cargoes could be landed or loaded only under the watchful eyes of customs officials in specified locations, the so-called 'legal quays' within the City of London (Figs 2.2 and 2.3). From henceforth, it was to these quays in the Pool of London that all ship-borne trade was off-loaded, save for sea-coal from Newcastle, beer and corn. The wharfage thus designated in the Pool of London was less than 600 m in length. Before 1559, although only Queenhithe, Dowgate and Billingsgate were the official legal quays, many wharves were actually used, not just on the London waterfront, but also at Gravesend, Woolwich, Barking, Greenwich, Deptford, Blackwall, Limehouse, Ratcliffe, Wapping, St Katherine's and Southwark. However, the new Act declared that:

> all and singular quays wharves and other places within the City of London and the suburbs (the

> several keys wharves and stairs and places before limited and appointed only excepted) ... shall be from henceforth no more used as landing or discharging places for merchandise but be utterly debarred and abolished from the same for ever (Dietz, 1972, 160).

The long and detailed recital of the names of the now illegal wharves is telling. Clearly, the early 16th-century port was deemed to have grown too large for the officials to control and thus was cut down to a more manageable size. This was a fulfilment of the age-old maxim that a port will develop not where it is most convenient for ships to berth, but where it is most convenient for customs officials to collect taxes; the location of a port is dictated as much by the prevailing administration as the prevailing wind. The impact of this radical re-focusing of the port's profits on the warehouses and wharfingers in Thames Street would have been met with delight by the controlling interest in the city as much as by the customs officials themselves. As for those maritime communities to the east of the Tower, whose livelihoods had just been erased, little is recorded, but much can be surmised.

Given the relatively modest extent of the legal quays, it is perhaps unsurprising to learn that an additional royal regulation was passed in 1564, decreeing that all cloth being shipped outwards should do so from Custom House Quay, while all imported fine and haberdashery wares must also be brought to the same quay by lighter. That these demands were being widely disregarded is clear from complaints made to the Privy Council in 1582 by George Needham, the Queen's Farmer of Custom House Quay. It seems that the influential companies, including the Merchant Adventurers, saw the system of restricted quays as inefficient and bad for trade. There were not enough lighters to move the cargoes quickly to and from the ships moored at Deptford, Limehouse or Blackwall to the Pool of London. Vessels were delayed and the goods themselves were spoiled by the weather as they lay on the quayside (Croft, 1973, xxvi). A compromise was reached, whereby a range of goods could now be shipped out from other quays 'at the merchant's pleasure' (Croft, 1973, xxvii). Thus some twenty years after the quays east of the City that had been 'utterly debarred and abolished forever' from the lucrative export trade, they were back in business; the prospering port needed to expand.

As for London and its new legal quays, these can now be reviewed, working with John Stow's late 16th-century *Survey* (Wheatley, 1956), the detailed documentary studies of Brian Dietz (1972, 156–67) and others, augmented where relevant by archaeological investigations and illustrated by contemporary panoramas, such as that by Antonis van den Wingaerde, (Colvin and Foister, 1996). From the landward side, the wharves were approached via the steep and narrow lanes that led from Eastcheap to Thames Street. The more level waterfront terrace that extended southwards from that busy east-west thoroughfare lay over land incrementally won from the river since the 11th century (Milne, 2003, 18–20). From the river access to the wharves would be from the private stairs or jetties that projected out over the foreshore or from the public water gates, such as those either side of Wool Quay at Water Lane and the Tower. The largest public water gate, and the therefore the most frequented, was at Billingsgate with a fourth river stair at Botolph's Gate and a fifth next to London Bridge itself (Wheatley, 1956, 40–1). Accessibility to the common lanes that led from the public water gates to Thames Street was a frequent source of complaint. The long-cherished rights of public access to the river (as opposed to the lading and unlading of customable goods) were as important a feature of civic concern in the 16th as in the 13th century (Loengard, 1989), there being no Thames Path or riverside Embankment in the Tudor harbour. It was the privately owned working wharves that occupied the majority of the waterfront sites in the Pool of London, a close-packed façade of two or three-story buildings, the skyline punctuated by the towers of the many churches that represented both the piety and the prosperity of the Protestant port (Brooke and Kier, 1975, 122–48; Wheatley, 1956, 434–7).

The closest legal quay to the Tower itself was **Galley Quay** (Fig. 2.3) in the parish of All Hallows, Barking. In the mid-16th century, it was described as 'greatly occupied with all kinds of merchandise' and was provided with a crane and a jibbet. **Old Wool Quay** had only been used for wool and felt, but more recently for wood and other aspects of the coastal trade (Dietz, 1972, 160–7). This site, at the extreme eastern edge of the city, had once been the heart of a major medieval shipyard, as recent excavations there have graphically confirmed (Damian Goodburn, pers.comm.).

Then there was **Custom House Quay** with its eponymous new brick building (Figs 2.2 and 2.3). The surviving illustrations of this building conflict in the information they present (Jarvis, 1976, 276), but the new complex may have incorporated a principal range of three storeys, perhaps with an open arcade at quay level with subsidiary ranges to east and west. It was here that the senior officials would be based, the collectors, the controllers, the head searcher, the surveyor and the tide waiters. The latter were the team who actually boarded the ships to determine the nature of the cargo carried, as the tide swept the vessels into the port to anchor. The petty customs accounts compiled by those officials demonstrate the widely fluctuating nature of London's trade in the mid-16th century: for example the figures for the five years following the reform are 1558/9 £30,951; 1559/60 £36,982; 1560/1 £27,605; 1561/2 £25,250; 1562/3 £18,146 (Dietz, 1972, xvi).

As for the Elizabethan Custom House itself, it was destroyed in the Great Fire of 1666 and a new enlarged building was erected over the site. Excavations here in 1973 (Tatton Brown, 1974; 1975) may have recovered the plan of its deep foundations, which seem to have been incorporated in the larger late 17th-century building that replaced it. The ruins of the Elizabethan Custom House were surveyed after the Great Fire, and found to be 33 ft

6 ins (10.2 m) wide. The new late 17th-century Custom House was 33 ft 10 ins (10.3 m) wide. This evidence strongly suggests that both buildings shared the same footprint and therefore the same foundations (Jarvis, 1976; Reddaway, 1940, 260–61). Although we have next to no documentary records to help us reconstruct its façade, there are a number of broadly contemporary illustrations of this part of the waterfront drawn in the late 16th and early 17th centuries in which the building is shown, usually as an incidental detail rather than centre stage – illustrations by 'Agas' (1562), Norden (1593), Visscher (1616), Gottfried (1638), Hollar (1647 – Fig. 2.2), Howlett (1663). As Jarvis points out, the images are at best confusing. His preferred assessment of the potentially conflicting pictorial evidence suggests that there was a main range of three storeys in brick with some arcading at ground level and octagonal stair turrets on the southern facade. There were also flanking buildings, rather than integrated ranges, possibly to the east and to the west. Such a complex may represent developments of several phases and even the main range may have been heightened or lengthened during its 100-year life (Jarvis, 1976). In broad architectural terms at least the façade of our red-brick building would presumably have shared elements of the Elizabethan Royal Exchange building (Saunders, 1997) and Archbishop Morton's gatehouse at Lambeth Palace (Tatton Brown, 2000, 53–4).

The structure of Her Majesty's two main wharves at Old Wool Quay and at Custom House Quay came in for criticism in contemporary surveys, since the former was an ageing timber structure badly in need of repair, while the latter, although faced in masonry, was now set back so far from the deep water channel at low tide that the lighters had difficulty berthing there (Dietz, 1972: xiii). The size of the two wharves were measured up at 91 ft (28 m) and 107 ft (33 m) in length respectively. The latter was some 48 ft 6 ins (15 m) broad (N–S) and had two jibbets; it was 'greatly occupied with all kinds of merchandise both inwards and outwards'. It was to here that all 'fine wares' and haberdashery had to be brought by lighter: indeed during this period it was still standard practice to offload all cargoes carried on large vessels onto lighters in mid-stream and then trans-ship the cargoes from the lighter to the wharf. The only exception to this labour-intensive system of double-handling was 'massy wares', material too heavy to handle in midstream. For those, the ship had to be brought alongside the quay and its cargo craned out; such a machine (powered by human energy) is shown on a sketch made by Alessandro Magno in 1562 (Weinstein, 1994, 35; Barron *et al.*, 1983).

Moving eastwards from the Elizabethan Custom House, the next series of wharves (Fig. 2.3) all lie beneath what is now the elegant Georgian Custom House of 1817–25, in the parish of St Dunstan-in-the-East. There was some archaeological observation on this extensive site during the excavation of its basements and foundations from 1813–17 (Milne, 2003, 11). Sixteenth-century surveys

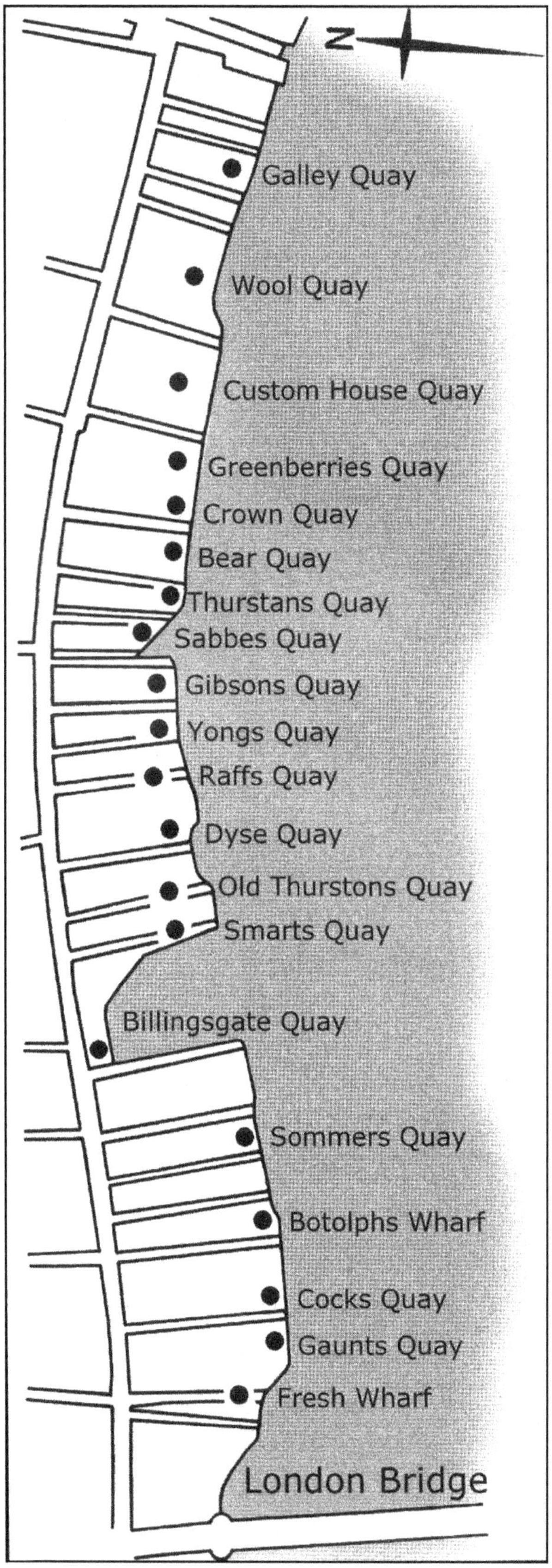

Fig. 2.3: The legal quays; plan view (north to the left) showing the location of the quays specified in the 1559 port reforms (drawn by Jenny Stripe, after Dietz, 1972, 166–7 – London Metropolitan Archives, City of London)

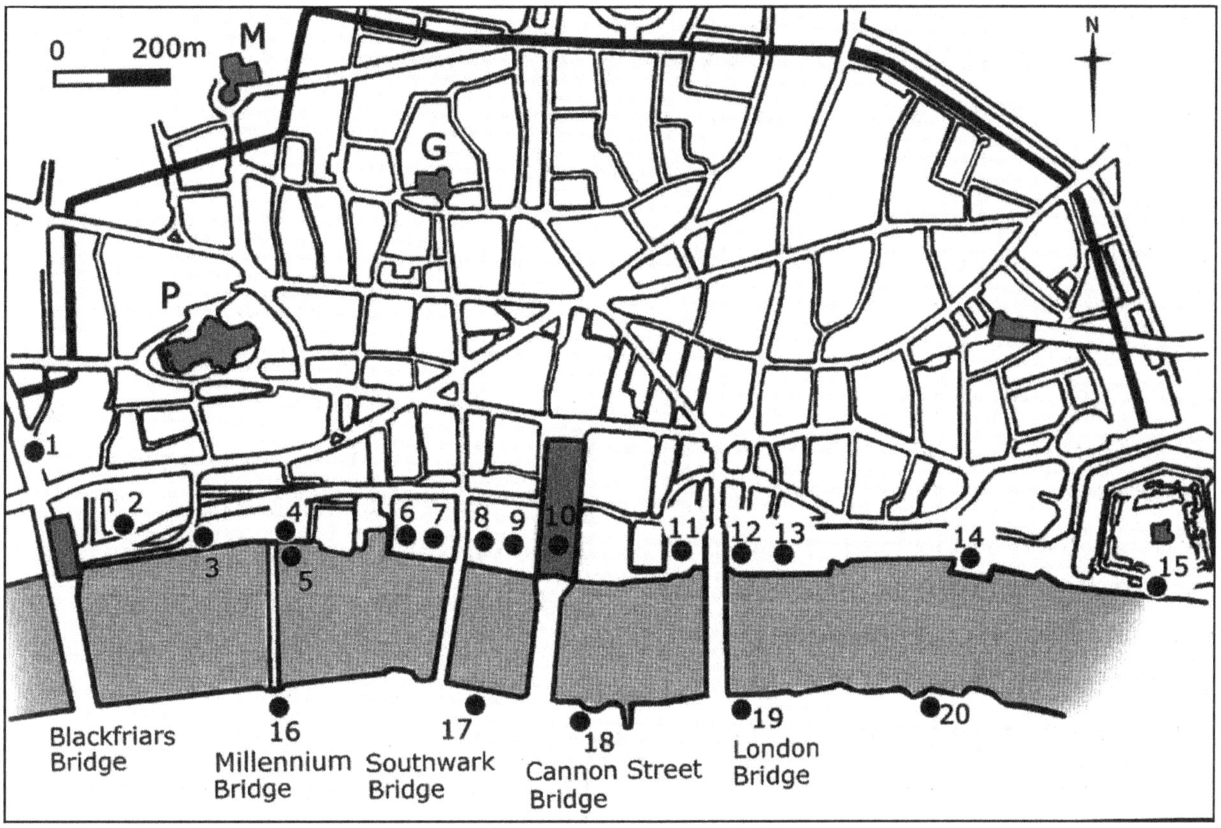

Fig. 2.4: London waterfront excavations: plan showing location of major riverside excavations of the medieval and early post-medieval harbour (drawn by Jenny Stripe – London Metropolitan Archives, City of London)

in 1559 and 1584 (Dietz, 1972, 156–60; 160–1) record that this was the site of **Greenberries Quay**, involved in the French trade, **Crown Quay** with the coastal traffic in corn and wood and **Bear Quay** with the Portuguese trade, while **Thurstans Quay** dealt with all manner of imports and exports. **Sabbes Quay** had a defective jibbet and dealt with 'pitche tar sope ashes and such like' and **Gibsons Quay** with lead, tin and other coastal trade. **Youngs Quay** accommodated Portuguese cloth merchants and **Raffs Quay** 'all kinds of merchandise inward and outward'. **Dyse Quay** was another of those used for coastal traffic, as was **Old Thurston's Quay**, but here they lay alongside Flemish traders.

The next set of wharves lay around the harbour at **Billingsgate** (Archer *et al.*, 1988, 53) in the parish of St Botolph and extended up to the Bridge itself. Starting in the east, was **Smart's Quay**, which dealt in fish, but according to William Fleetwood, the City Recorder in the 1570s, the wharf was noted for other activities. An alehouse here was a focus for organized crime, where a former merchant called Wotton had set up a school to train young boys to be cut-purses or pickpockets (Weinstein, 1994, 47). Adjoining it was **Sommer's Quay**, mainly used by Flemish merchants, while **Botolph's Wharf** lay next to the church and was used by coastal traffic, as well as by the Russia Company, who had warehouses for their wine there. Cargoes that would have been brought here on ships such as the *Swallow*, the *Lyon* and the *Charity* are listed in the 1567/8 Port Books and include many hundredweight of wax, tallow, tuns of fish or whale oil, cordage, tar, sealskins, wolf skins; wolverine, timber, mink, ermines, 100 beaver bellies; 5 beaver backs and 20 lbs (9 kg) of rhubarb. **Cock's Quay** was given over to foreign merchants who had storage facilities and lodgings there. Next came **Gaunt's Quay**, where barrels of fish were landed without use of a crane and **Fresh Wharf**, used by eel and fish suppliers (Dietz, 1972: 163). These wharves lay in the parish of St Magnus, the church that stood on the landward end of London Bridge.

Again, there have been a series of archaeological excavations on the sites of some of these quays (Fig. 2.4), notably, Botolph's Wharf, Gaunt's Quay, Cock's Quay and Fresh Wharf (from 1973 to 1983 see Steedman *et al.*, 1992; Schofield, 1994, 221–3; Schofield, 2013, fig. 5). The excavations of 1982–3 at Botolph's Wharf, for example, demonstrated a remarkable preservation of medieval buildings. They had been protected beneath a 'blanket' of debris spread over the legal quays after the Great Fire of 1666 to raise the general level above that of the mean spring tides. All the buildings excavated here are late medieval in date and survived largely unchanged into the late 17th century. Although the new legal quays might have had a brand new Custom House, the rest of the port seems to have remained in private hands, still largely medieval in form, fabric and scale. The highly sub-divided medieval tenement layout is not the layout of large-scale purpose-built warehouses. It would seem that the London properties were rented out to whosoever needed the storage space on short- or long-term leases. Little major 16th-century investment in infrastructure was apparent in the 100–200-year old medieval buildings that still conformed to a 400–500-year old tenement pattern.

One of the two principal Fish Markets was held in Billingsgate, around New Fish Street Hill. In 1561 a decree was issued to promote the eating of fish, not just on Fridays, but also on Wednesdays. As befits the mid-16th century, this move was not fuelled solely by religious fervour, but by the pragmatic need to revitalize and expand England's maritime industries. Such a move would see more ships built and would keep her mariners gainfully employed, an essential requirement 'should the necessity of war break upon us'. The London fishmongers warmly welcomed the initiative, whatever its motivation; the fish markets at Billingsgate and at Old Fish Street Hill continued to thrive as a consequence (Archer *et al.*, 1988, 55; 73).

There were just two principal wharves above the Bridge in 1559. First there was the **Steelyard**, where the Hanseatic merchants had their hall and wharf for their own exclusive use, a tenure that lasted from at least the 12th century. Although it had been seized by Edward VI, its privileges had been restored when Mary Tudor came to the throne in 1553, but the alien enclave finally fell in the last decade of the 16th century. The significance of the role Hanseatic merchants played in Tudor trade is shown by the large proportion of German stoneware in the total of all London's imported pottery from 1480–1600. Just to the east of the Steelyard was **Three Cranes Wharf** in the Vintry, the great wine wharf. Again both sites have been subject to archaeological investigations in the 1980s, revealing much about their ancient medieval ancestry (Milne, 2003, 23–4). An important market for grain and foodstuffs for local consumption was maintained at Queenhithe, set just to the west of the Vintry (Archer *et al.*, 1988, 75).

It will have been noted that Southwark, on the southern shore opposite the legal quays, was not seen as part of the formal port, since it lay outside the jurisdiction of the city fathers. This borough is remembered today for its association with Shakespeare and Marlowe through its famous playhouses, such as the Rose, the Swan and the Globe, all of which were built between 1587 and 1599 (Bowsher, 1998, 12–13) during the operative life of the Gresham Ship. But, with regard to that heritage of high culture, it should be remembered that substantial remains of a bear-baiting pit were recently excavated in the aptly named Bear Gardens nearby, complete with an assemblage of large dog bones (Mackinder and Blatherwick, 2000). Indeed, the Bankside area was better known in the 16th century, not just for its theatres, but for its stews and brothels, all of which would have added colour to the life of the harbour, whether by day or night (Carlin, 1996). John Stow notes that the English 'disdained to be bawdes' and that women from Flanders came over 'for that purpose' (Wheatley, 1956, 361). Based on an account provided by ancient men 'of good credit', he also learned that the single women who plied their ancient profession there were forbidden the rites of the church and a full Christian burial 'if they were not reconciled before their death' (Wheatley, 1956, 361–2); apparently their male customers had a better chance of salvation.

Harbour Construction

Although the cladding of London's waterfront had been in timber for most of the medieval period, the change to more durable masonry river walls had begun by the 14th and early 15th centuries (Milne, 2003, 142–4). Examples have been recorded at sites near the Millennium Bridge, Baynard's Castle and are still extant at Tower Wharf itself and at the Jewel Tower, Westminster. The Tudor river wall at Whitehall was recorded in a pioneering archaeological investigation that commenced in 1939 (Thurley, 1999, 52–9). However, not all riverside properties were stone-faced in the 16th century, as the City Viewers' records show. These officials inspected sundry building complaints in the Tudor town, including several down on the waterfront (Loengard, 1989). In 1535, for example, we read that Marlowe's Quay required carpenters for work on the wharf and 'to amend the framework of the same towards the Thames' (Loengard, 1989, no. 115). The same was true of the jetties and stairs that projected out into midstream. There had been a dispute in 1526 over a new river stair built at Sabbes Quay, for example; the great timber frame extending 24 ft (7.3 m) southwards into the river, was 8 ft (2.4 m) broad and the stair itself descended some 19 ft (5.8 m) onto the foreshore at low tide. It was decided that the frame, bridge and stairs should be removed, since they were built on common ground 'to the great nuisance and hindrance ... as well of the King's subjects as all manner of ships, crayers and boats repairing to this city' (Loengard, 1989, no. 68). We also learn that the responsibility for maintaining the waterfront rested squarely on the shoulders of the landlord, not the City, as is shown by a case in the parish of All Hallows, Barking, in 1550; here a 25-ft (7.6-m) length of wharf was decaying, together with its house and stables. The lords of the house were requested to 'repair and amend them at their own proper costs and charges' (no. 285).

Ships, boats and barges (Fig. 2.5)

A contemporary description of the late Tudor harbour records a bustling river that, quite apart from the 'great ships and other vessels of burthen', was worked by some 2000 passenger wherries and other small boats 'whereby 3000 poor men, at the least, be set to work and maintained' (Wheatley, 1956, 13; Humpherus, 1859). The minimum dimensions for any vessel used for rowing or carrying passengers on the Thames must not be less than 'xxij foote & a half in lengthe, and fower foote—-- and a half broade in the mydshippe', i.e. 6.7 m by 1.4 m, according to mid-16th-century legislation (*The Statutes of the Realm*, 1819, **VI(I)**, 291). Elizabeth, wary of the threats posed by her neighbours across the Channel, regularly recorded the number of privately owned vessels that could be requisitioned for military service, should the need arise. Sometimes these records included the number of men available to man the ships; the national returns for 1582, for example, list nearly

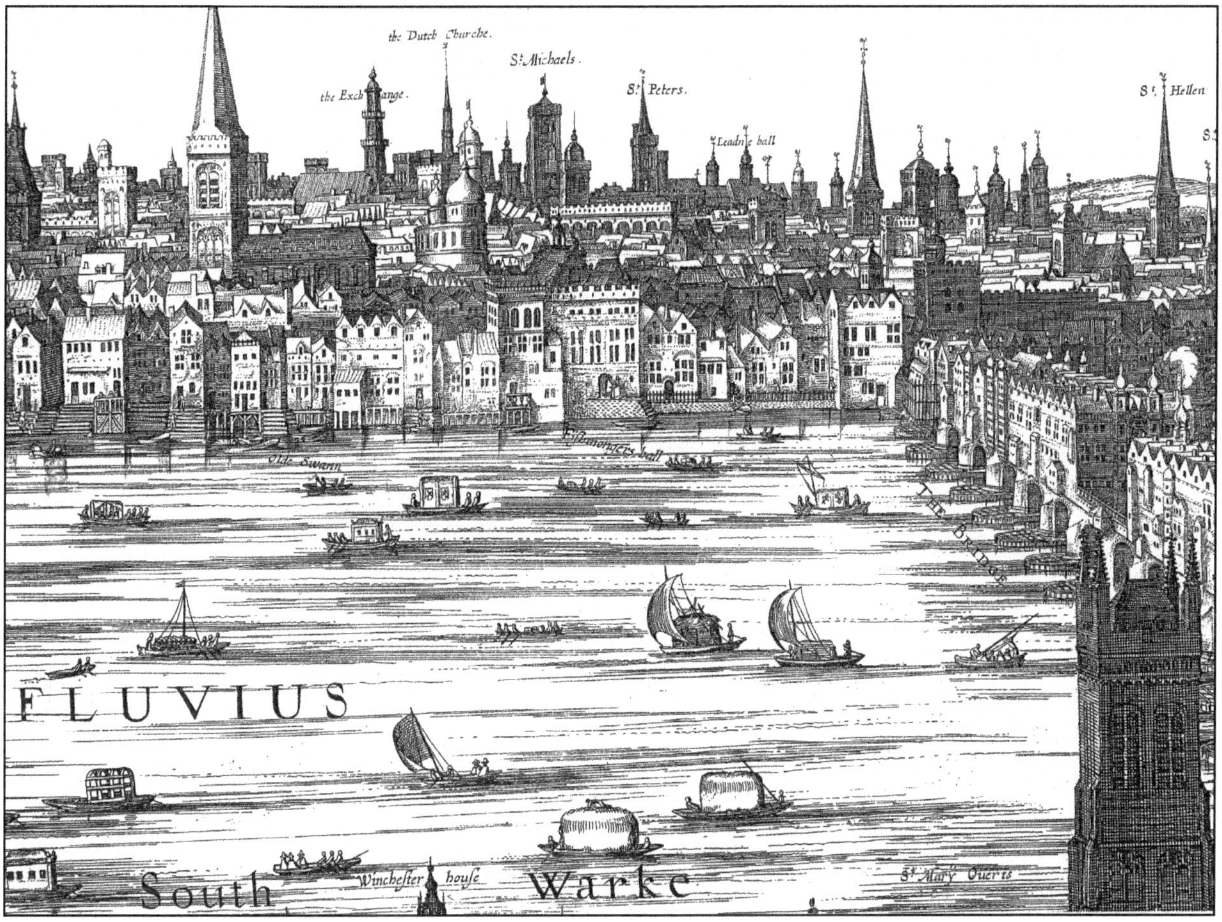

Fig. 2.5: *Boats and Barges on the Thames: detail from an early 17th-century panorama showing range of small boats and barges that worked the Thames to the west of London Bridge and the Legal Quays (C. Visscher* c. *1616, © London Metropolitan Archives, City of London).*

1500 masters, 11,500 seamen and over 2000 fishermen. To that naval reserve were added nearly 1000 London watermen, all expert oarsmen (Oppenheim, 1896, 177). There is archaeological evidence for the river craft they would have employed, in the form of fragments of a number of 16th-century boats reused as revetment cladding or lining drains recovered from waterfront sites in London. They all represent clinker-built boats, in which the overlapping strakes were clenched together with iron nails and roves. Such a technique was once also utilized in larger ship construction, but by the mid-16th century had long been superseded by the carvel technique, as exemplified the Gresham Ship. The London boat finds therefore represent not just a particular class of vessel, but show that boatwrights and shipwrights were now two distinct trades, operating on different sites, with different tools and requiring different resources. A ship was no longer a scaled-up version of a river boat, as had been the case in the 14th or 15th century (Milne, 2003, 165–74).

The 16th-century boat fragments recorded from London waterfront sites have been catalogued by Peter Marsden (1996, 218–21, nos 3; 4; 7; 8; 9; 10; 14; 18; 19; 20; 23; 38), their ubiquity on modern London waterfront sites reflecting their dominant presence on the Tudor river. There were ten times as many oared vessels on the Thames as sailing ships. In addition to these vessels, evidence of larger Thames working barges or lighters has also been recorded. Some of these craft were also constructed in the clinker-tradition, with a much repaired 17th-century lighter – Blackfriars 2 (Marsden, 1996, 154–9) – showing some constructional similarities with two working vessels of 15th-century date – Blackfriars 3 and 4 (Marsden, 1996, 55–106). The Tudor barge traffic may thus have had more in common structurally with the smaller wherry boats than with the larger sea-going vessels they worked alongside. That said, the more recent discovery of a 6-m section of a particularly robust late 16th-century river barge shows a quite different construction technique (Goodburn, 2002, 289–97). The vessel may have been some 12 m long, a beam of 2.5 m and a depth of 1 m and was probably constructed *c.* 1580. This carvel-built vessel was a large punt-like (swim-headed) flat-bottomed craft, perhaps one of the so-called 'western barges' represented on 16th- and 17th-century panoramas of the working river (Sturdy, 1977, fig. 22.3; Wilson, 1987, 35–42).

As for the 'great ships and other vessels of burthen', there is also some archaeological evidence for them from the Thames. The substantial remains of a ship believed by some to be the *Royal Sovereign* was recorded in 1912 at Woolwich (Marsden, 1996, 212), while the Gresham Ship was recovered from further out in the estuary a century later. The structure, form and performance of this merchantman, together with comparisons with other contemporary vessels, are discussed in detail in the first volume of this monograph by Jens Auer and Thijs Maarleveld and their team.

Shipping in the port

A survey of English shipping compiled in 1582 recorded that there were 128 ships registered in London, with a total tonnage of 12,708, and some 2286 associated crew members. The totals for this one port are as large as those listed for whole counties (for example Suffolk, Essex and Sussex), highlighting the dominant nature of London in this period (Friel, 2003, 112–14). Any more detailed analysis of the shipping in the port must start with study of the Port Books. In November 1564 William Paulet, Marquis of Winchester and Lord High Treasurer of England, issued instructions that required custom officials to fill in annual returns listing all the ships and customable cargoes that arrived in the port (Williams, N., 1955). Unfortunately only two of the London Port Books survive from the Elizabethan period and of these only one has been transcribed and edited in an accessible format. This is the return for 1567/8, published by Professor Brian Dietz (1972), an essential foundation for any assessment of the Elizabethan port. The Port Book is a catalogue of 830 ships and their cargoes that the customs officers examined, quantified and costed to establish what taxes could be extracted. It was initially compiled by the Controller, the aptly-named William Temys, and his clerk, working methodically and conscientiously (Dietz, 1972, xvii-xx). The name of each ship (variously spelt), its tonnage, its master, its assumed home port (for example 'of London') and its most recent point of departure (for example 'from Bordeaux') are listed (see Table 2.1). The cargo (or at least those elements for which duty was payable) are then enumerated by tying together the captain's documentation with what was actually seen by the Searchers from the Custom House as the cargo was being transferred from the hold of the ship to lighters or as subsequently exposed on the quayside. Each day's listings were entered into a Port Book and duly dated.

This list is by no means a complete survey of every ship in the port. Not every ship that arrived in London is mentioned; neither are the cargoes for which no tax was required. The year 1567/8 (Michaelmas to Michaelmas – 29 September) is one of only two years for which such a complete record survives from this period and thus detailed annual comparisons are not possible. All that said, there is still much to learn from these records, which most certainly illuminate the workings of the port. The entries show that the majority of the port's contacts in this particular year were relatively limited, extending in the main from the Zuiderzee to Marbella. Indeed, the trade to and from the Netherlands dominates the picture, with, for example, 117 arrivals from Antwerp and 95 from Amsterdam. By the end of that year, politics intervened and the long-lived lucrative Antwerp axis was closed to English merchants. This had happened before (Alsop, 1996, 154), but this time there was no return: the Merchant Adventurers were obliged to move their Staple to Hamburg in 1569. By the end of the century London's enterprising shipping had been forced to look further afield.

Extending London's mercantile reach in such an age was not without risk; the trade with Spain can stand as an example. From at least 1530 a group of English merchants trading in the Iberian Peninsula in the key port of San Lucar established a 'Brotherhood of St George'. From these origins, the Spanish Company was formed, with a Royal Charter signed in 1577; among honorary members listed in that document were Walsingham and Leicester, as well as Sir Thomas Gresham (Croft, 1973, xiv). In spite of such influential patronage, success was not assured, for that same year Drake began his campaign attacking Spanish shipping in the New World. This had serious repercussions for English shipping harbouring in Spain and for Englishmen living there. Delicate negotiations were the order of the day, not alas assisted by the new religious thinking. Some of the merchants working in Spain had Spanish wives and remained Catholic; the newer breed of English entrepreneurs were hard-line Protestants, while some preferred to be Catholics abroad and Protestants at home (Croft, 1973, xxi). But by 1586 diplomacy was finished; their vessels were requisitioned to serve in the Armada and the men were imprisoned to await the attentions of the Inquisition (Croft, 1973, xxviii).

The profound change in London's horizons was assisted by a new generation of seafarers and vessels such as the Gresham Ship. In 1567/8 some 736 ships with a total tonnage of 34,600 arrived in the port of London, but by 1600, those figures had increased to 1037 ships, with a combined tonnage of 52,300. In the late 16th century, the Pool of London had therefore become far more crowded. The volume of traffic certainly increased dramatically by 1600, but there is still a familiarity about the content of the trade itself. Ships dealing with the southern European trade had seen a modest increase from 84 to 92. Those trading with northern Europe had almost doubled from 40 to 77, but by far the largest increase in traffic was with our near neighbours in north-west Europe. In 1567/8 London's trade with this sector involved 612 ships; by 1600, that figure had increased to 864 (Dietz, 1986, 128).

But just looking at the native shipping is also informative. In 1567/8 there were 300 English vessels in the port, with a total tons burthen of 16,800; in 1580/1, this had increased to 488 and 30,881 tons. In 1587/8, the year of the Armada, this fell back to 305, totalling 21,919 tons, increasing to 340 and 26,941 tons the following year. A decade later, in 1599/1600, the figures were 314 English ships, totalling some 27,000 tons (Dietz, 1986: 128). There are therefore marked fluctuations in the quantity of traffic, but one trend is constant and that is that the average size of the English vessels in the port increased over this period. The average in 1567/8 was a modest 56 tons; in 1580/1 it was 66 tons, in 1587/8, about 70 tons, in 1589/90, about 80 tons and by 1600 it has risen to 85 tons. Once again, the Gresham Ship must be set against this background, for it can be seen as larger

Port of Elizabethan London
Ships and their Masters 1567/8

Name	Tonnage	Master	Dietz 1972 – Calendar No
Abram	40	Peter Marram	26
Anne Gallant	45	Arthur Pitts	353
Anne Sackford	55	John Renold	286
Audry James		John Burrell	659
Awdry Joanes	70	John Burwell	112
Bark Aldersey	70	Thomas Rickman	592
Bark Allen	80	Lawrence Hayward	78
Bark Allen	80	John Pearson	415
Bark Bond	70	John Benet	204/551
Bark Collet	120	Jacob Riddam	679
Bark Fox	50	Thomas Bolton	830
Bark Fox	50	Thomas Pirrie	272
Bark Gray	70	Robert Gray	197/593
Barnaby	45	Peter Hills	236
Barthelemew	60	Richard Cokes	627
Beniamen	60	John Dryver	790
Beniamin	70	John Dunton	88
Bonadventure	70	Thomas Lawson	230
Charitie	100	Thomas Awdley	805
Charitie	100	John Webb	472
Charity	130	Richard Gybbes	40
Christ	90	Bartholomew Storm	246
Christopher		William Langley	360
Christopher	60	William Langley	677
Christopher		Thomas Wilson	514/682
Clement	70	Thomas Rickman	95
Confidence	50	Richard Cornish	1
Confidence	48/45	Erasmus Kidball	450/641
Confydence	60	Erasmus Kedball	256
Edward	50	William Cox	789
Edward	40	William Holland	248
Edward	50	Christopher Hubbord	755
Edward	40	Andrew Luce	64
Elizabeth	80/60	Robert Burrell	81/443
Elizabeth	50	Thomas Harrison	680
Elizabeth	80	Thomas Sherland	242
Elizabeth	40	Edward Smarley	247
Elizabeth	40	William Tyllor	760
Ellen	80/60	William Spender	298/576
Falcon		Dignatus Garlof	510
Falkon		Thomas Grene	613
George	60	William Thomson	356
Grace of God	30	John Stephen	38
Grace of God	45	Jacob Steward	191
Grace of God	45	Richard Tye	380/827
Grace of God	50	Thomas Wilson	203/631
Hart	110	Thomas King	558
Hearn	40	John Davis	636

Table 2.1: Port of Elizabethan London: Ships and their Masters 1567/8: Ship names recorded in the Port Book returns 1567–8 for vessels listed as being 'of London'. Each entry (catalogue no.) represents a different dated event; where the same name is listed, the recorded tonnage or the Master's name may differ (after Dietz 1972)

Name	Tonnage	Master	Dietz 1972 – Calendar No
Jesus	20	Thomas Clerk	253
Jesus	70/80	John Clerk	339/526
John	40	Stephen Read	710
John	30	John Sherwood	3
John	40	Thomas Turner	456/759
John Baptiste	70	William Hall	341
John Baptiste	65	Thomas Perry	754
John Baptiste	45	Robert Temple	428
John Baptiste	50	Thomas Wright	207
John Bonadventure	60	Richard Say	522
Julyan	130	William Oxley	245
Julyan	90	William Uxley	784
Katharine	80	Thomas Pollard	685
Lion/Lyon	80	Richard Bebell	77/331
Lyon	75	John Dunstan	715
Marygold	45	Thomas Andros	416/ 556
Mary An	50	Thomas Dale	141
Mary Anne	80	John Bonner	143
Mary Anne	60	Thomas Dale	425
Mary Edward	70/60	Edward Master/Maister	289/527
Mary Fortune	80	Thomas Lawson	233
Mary fortune	28	Richard Mills	538
Mary Fortune	70	John Morcok	584
Mary George	70	Jacob Fryer	644
Mary George	80/80	John Fryer	250/714
Mary George	45/50	Robert Osburn	2/767/335/555
Mary George	100	Peter Richardson	703
Mary Jermaine	50	William Crag	135
Mary Katharine	100	William Thomas	809
Mary Martin		John Ewstas	330
Mary Thomas	50	William Dixson	226
Mary Thomas	25	William Dixson	751
Mary & John	55/50	John Diet/ Dyet	66/418/769
Maudlen	50	Andrew Markes	290
Nightingale	30	John Thorneborrow	221
Pellican	150	Stephen Johnson	728
Pellican	40	Harry/Henry Wright	102/553
Pellican	50	Richard Guydy	198/ 681
Pellican	30	James Steward	761
Peter	60	Richard Clark	741
Peter	55	Randolph/Ralph Goodwyn	127/449
Richard Arundell	90	Richard Twid	689
Saviour	80	Dany Phillie	268
Star		John Podge	4/686
Swallow	120	Stephen Aborro	39
Swallow		Giles Gray	435
Swallow	60	Richard Poulter	795
Thomas	40	Robert Smyth	37
Thomas Allen	140	Thomas Wade	332
Trinetie	55	John Dryver	124
Trinity Richard	58	Richard Clerk	453
Unicorn	45	Jacob Cornilison	52
William	60	Thomas Foster	202/515
William Bonadventure	80	Thomas Lawson	525
Willeby	120	Arthur Petts	785

Table 2.1 (cont.)

than average for the port, clearly a long-distance merchantman.

In sum the record on 1567/8 should not be seen as a typical year and thus must be treated with due caution. But such caveats aside, much can be learned from the detailed 1567/8 Port Book returns, not just about the levels of tax raised or the efficiency of the customs officials, as Brian Dietz has shown so impressively (1972, ix–xx), but also about the vast range and scale of the imports arriving in London; an alphabetical commodity list runs to fourteen pages (Dietz, 1972, 138–51). Of particular interest to the Gresham Ship team, for example, was the mention of iron being brought to London in English merchantmen, presumably for re-export. Three of these ships came from Antwerp: the *Primrose* with 15 cwt (762 kg) of 'ames' iron (perhaps from Amiens – Dietz, 1972, no. 45), the *Lion* with 26 cwt (1320 kg) steel and 1 ton 13 cwt (1676 kg) of small square iron (Dietz, 1972, no. 18) and the *Hound* with 10 hd iron plates and 25 cwt (1270 kg) of "Lokes" iron (probably from Liege – Dietz, 1972, no. 466). Two vessels came from Gdansk: the *Nightingale* with 1 cwt (50 kg) of "spruce" iron (Dietz, 1972, no. 145) and the *Jonas* with 5 tons (5.1 tonnes) of Spanish iron and 5 tons (5.1 tonnes) of faggot iron (Dietz, 1972, no. 557), while the *Falcon* brought back 50 tons (50.8 tonnes) of iron from Bilbao (Dietz, 1972, no. 63). All these consignments were parts of very mixed cargoes; there is no evidence for a merchantman carrying just iron or even just a range of heavy metal bars and ingots. Just how mixed Elizabethan cargoes were can be shown, by way of example, through the Port Book entry for the consignment carried by the *Hound,* mentioned above. In addition to the Lieges iron, there were hops, woad, madder, many carpets and textiles (including taffeta and lace), canvas, aniseed, ginger, pepper, 1000 pins, crossbow thread, 21 cwt (1067 kg) of frying pans, 3 gross (432) harp strings, eight Cologne lutes and drugs *ad valorem* £4 (Dietz, 1972, no. 466). Had that ship sunk in the estuary, it is worth pondering just how much of such a cargo would have survived on a wreck site for four hundred years.

The Port Books also reveal much about a working port, its modus operandi and its annual rhythm, features which may have remained much the same, even though the later ships may have been larger and their cargoes derived from a somewhat wider world. The dated entries made by the customs officials in 1567/8 record the day a vessel's presence was noted in the port and the initial assessment of the cargo made. For some vessels, such as those carrying a broadly uniform cargo of wine in casks systematically and symmetrically stowed, it was just a matter of counting and checking the master's documentation with the actuality. This paperwork seems to have been completed for most of the Bordeaux wine fleet that arrived in November 1567 in one or two days; presumably the task of delivering these tuns to the appropriate waterfront warehouse took rather longer.

For some of the larger merchantmen engaged in other trades, the Port Books record not just the arrival date, but further dates are added in tandem with the continuing examination of complex mixed cargoes from the holds. This process of carefully checking would consequently take as long as it took for the hold to be emptied by hand and crane (Williams, N., 1955, 16). We therefore have an indication of the varying timescales of such operations. By way of example, the 200-ton *St George* from Gdansk was first noted in the Port Book on 16 June 1568 (Dietz, 1972, no. 567), when five lasts of pitch were recorded. Consignments of flax, more pitch and 480 soap boxes were counted on 17 June; 86 cwt (4.4 tonnes) of cordage on 18 June, but the rest of the cargo, including 3 vats of eels and four tons (4.1 tonnes) of iron were not finally logged until 5 July. As for the larger 350-ton *Santa Maria de Gracia* from Venice (Dietz, 1972, no. 53), work seems to have begun unloading 268 butts of wine on 22 October 1567, but the officials were still counting 103 more butts on 27 October, 64 lbs (29 kg) of marmalade and 16 lbs (7.3 kg) of green ginger on 4 November, before the final entries were made dated to the 14th of that month. Such large merchantmen could thus expect to wait at least two to three weeks before a full cargo could be transferred and presumably a similar length of time before an export cargo could be loaded. Thus an approximate turn-around time for these Elizabethan ships can be surmised.

That said, nearly 250 of the 830 vessels listed in the Port Book were able to make at least two return journeys within the twelve month period recorded. Several managed three round trips (e.g. the *Bear* from Amsterdam and the *Christopher* from Antwerp). Eleven managed four visits (for example, the *Cock* from Bruges, the *Grace of God* from Leigh) and some managed five visits (e.g. the *Jonas* from Flushing), while the *Sampson* from Bruges and the *Spreadeagle* from Dordrecht logged seven visits each. Such a regularity of contact must have bred a strong sense of familiarity amongst all parties, the customs officials, the London merchants and wharfingers as much as the mariners themselves. It would also have provided regular opportunities to observe at close quarters the ships built in different yards in different traditions, from the Low Countries, from Germany, from Scandinavia, from France, from the Iberian Peninsula and even from Venice (Fig. 2.6). A busy international port would serve as an informative forum for exchanging ideas on ship-building as well as on more mercantile matters.

The traffic in and out of the London harbour does show great variability, however. November 1567 saw the port at its most busy, with 125 entries, boosted by the Bordeaux wine fleet; July and August were also busy months with over 90 ships. In stark contrast, January and February saw just 54 and 33 arrivals respectively. The fluctuations were just as dramatic on a weekly or daily basis: On 6 November 1567 the arrival of just one new vessel is recorded; on 12 November a fleet of seventeen

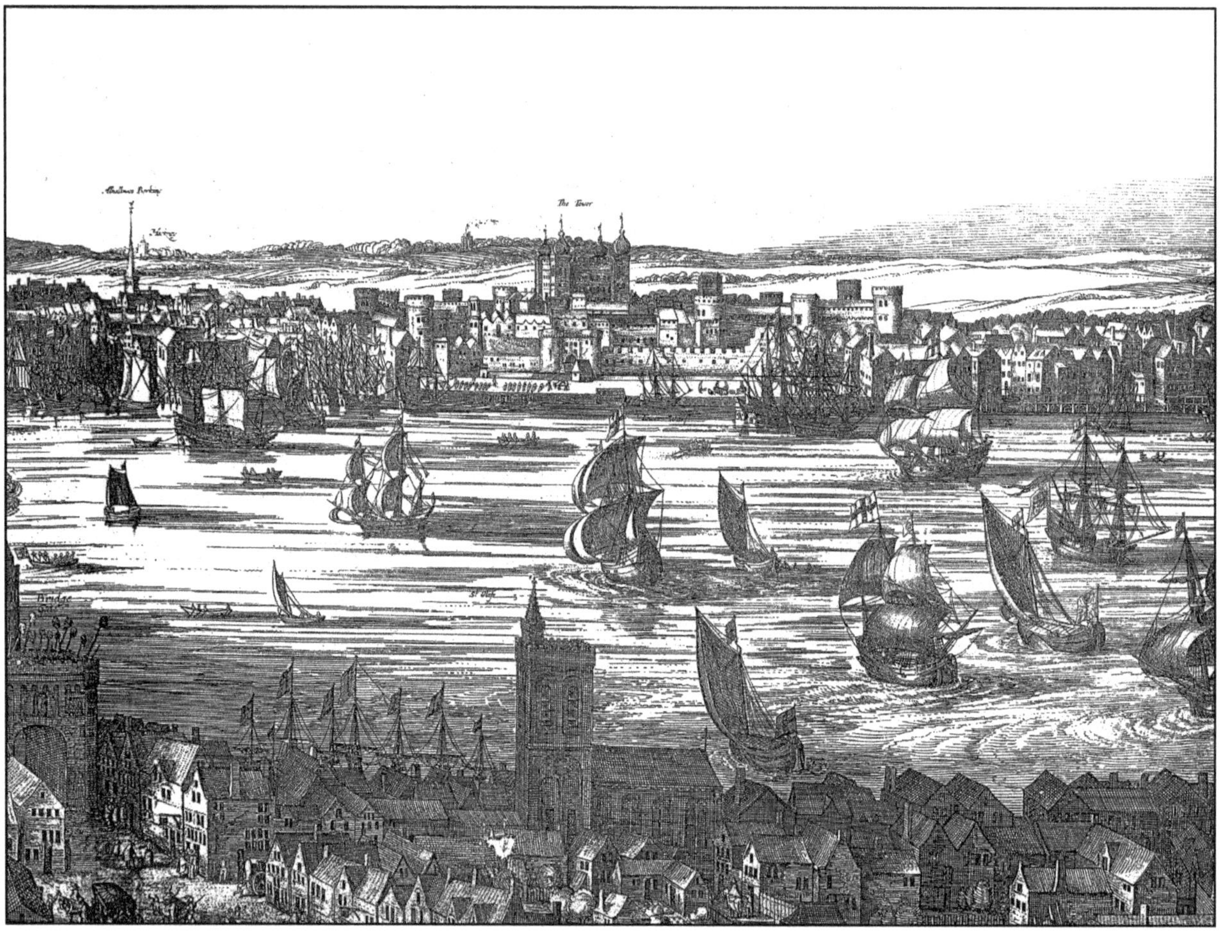

Figure 2.6 *Many Greete Shippes: detail from an early 17th-century panorama showing range of larger, sea-going merchantmen that sailed into the busy harbour to the east of London Bridge (C. Visscher c. 1616, © London Metropolitan Archives, City of London)*

arrived. And there were days when no new vessels arrived at all; from 24 to 31 January 1568 just one vessel is logged as arriving in the port. Whether this reflects a harsh winter, contrary sailing conditions or a harsh economic climate is hard to be certain of, although the former options seem most likely in January.

The implications for those whose livelihood depended on off-loading, transferring and storing new cargoes are plain to see; some days, some weeks, some months (such as November) would be busier than others and sometimes there would be no work at all. Since Mayhew recorded his trenchant observations in 1851 (Quennell, 1984, 563–72), much has been written about the inequities of the casual labour system in 19th-century docklands, where men were called on for half a day's work as required and laid off as the ship left. How such an irregular work flow was managed in the Tudor port is less clear.

The working year (Table 2.2)

For those who were employed in the Protestant port, at least as customs officials, but presumably also as watermen, wharfingers and so forth, the week lasted from Monday to Saturday. According to the dated daily entries, no work was expected on Sundays, regardless of the tides, but a number of other days in the calendar are also omitted from the schedule. There are dates when not only are no new vessels noted in the harbour, but also when there are no entries concerning the protracted examination of the more complex mixed cargoes, as discussed above: in other words, weekdays when the Custom House was clearly not operative. These non-work days require explanation, since they appear every month. Severe inclement weather might account for some of the omissions, especially those in January. On a happier note, there are a number of high days, Holy Days and other celebrations which seem to coincide with some of the 'missing' days. However, the Port Books suggest that other well-known Tudor festivals including St Valentine's Day (14 February); All Fools Day (1 April), St Swithin's Day (15 July) and St Crispin's Day (25 October) were not recognized, at least not within the Custom House.

Palaces and people

The banks of the Thames were more than a working harbour: it was a favoured location not just for waterfront warehouses, but also for grand buildings, with the river providing a broad and convenient highway. At the eastern end of the City was the Tower, part palace, part prison, part armoury; at the western end was Baynards Castle, a 15th-century castellated mansion that passed through royal hands and where the last Plantagenet

A Year in the Life of the Elizabethan Port of London

October 1567:	Sixty-six ships recorded in the port. Discounting Sundays, there were four days with no corresponding notes in the Port Book, the 11th, the 18th, the 28th and the 29th. The last date was to celebrate the secular pageant of the Lord Mayor's Show, a festival established in 1546 to replace the quasi-religious Midsummer Watch (Weinstein, 1994, 54).
November 1567:	One hundred and twenty-five ships recorded (largest total of the year). Just one non-working day to account for, which is All Souls Day (the 1st)
December 1567:	Seventy-five ships recorded. The Christmas festival lasted from the 25th to the 27th.
January 1568:	Fifty-four ships recorded. Of the non-working days, the 6th January was Twelfth Night, but the 9th, the 24th to the 29th and the 31st might have represented a winter freeze.
February 1568:	Thirty-three ships recorded (the lowest total of the year). Of the non-working days, Candlemas – the first day of spring – was celebrated on the 2nd, but in spite of that the 7th, the 18th and the 25th might represent more poor weather.
March 1568:	Seventy-six ships recorded. The 2nd and the 25th were non-working days.
April 1568:	Sixty ships recorded. The Easter Festival is marked by a seven-day break (longer than Christmas) from Holy Wednesday (the 15th) through Easter Sunday (the 18th) to the 21st.
May 1568:	Fifty-two ships recorded. The merry month begins with the May Day festival (the 1st and the 2nd, including the Sunday). Saturday the 8th was also not worked; neither was Ascension Day, the 28th.
June 1568:	Fifty-seven ships recorded. Of the five non-working days, the 7th to the 9th might relate to Pentecost, with Midsummer's Day on the 24th, and then the 29th.
July 1568:	Ninety-two ships recorded. Two non-working days were the 21st and the 24th.
August 1568:	Ninety ships recorded. Of the four non-working days the 2nd was Lammas, or Loaf-mass Day, to celebrate the harvest festival. The other days were the 4th, the 5th and the 25th.
September 1568	Forty-four ships recorded, but the accounts for this month seem incomplete: the 17th and the 21st were non-working days, as was the 29th, Michaelmas Day, close to the equinox and end of the financial year.

Table 2.2: *A Year in the life of the Elizabethan Port; a month-by-month tally of the ever fluctuating number of ships recorded as arriving in the port. The harbour usually operated a six-day week, but this table also records days when the Custom House officials did not seem to be at work, either because of bad weather or for holidays/feast days (based on the London Port Book for 1567/8, Dietz, 1972)*

Richard III was offered the English crown. Excavations in the 1970s showed that the robust foundations of this large masonry building, like so much of the medieval City, survived the 16th century with little superficial change, only to be burnt out in the late 17th century (Dyson, 1989: 9–12; figs 14–18). The opposite bank of the Thames, in Southwark, also accommodated a number of fine palaces, including Rotherhithe Palace (Platform Wharf), dating from the mid-14th century, as well as Dunley's moated house, Edward II's Rosary, the 15th-century Fastolf Place, not to mention a tidal mill (Bluer, 1993; Blatherwick and Bluer, 2009). All these have been excavated, as has much of the Bishop of Winchester's Palace (Philpotts *et al.*, 2006) to the west of the Bridge.

The location in London of such buildings, together with the palaces of Westminster, Whitehall and the Archbishop of Canterbury's house at Lambeth, represent the social elite; the City's markets and craftsmen were working for the deepest pockets in the land and the port provided that favoured clientele with whatever their money required. The ready cash attracted the merchants; the rich markets attracted the discerning buyers. This was a virtuous circle of a size enjoyed by no other town in the land. Thus through this mutual interdependence the personal connections between the merchant class, the City Companies and the Court grew ever closer. Robert Cecil was Elizabeth's chief adviser on statesmanship and national expansion, while Thomas Gresham had an unequalled grasp of financial matters (Broodbank, I, 1921, 52). This was a world of patronage and obligation, relying as much on charisma, trustworthiness and reputation as on balance sheets and financial predictions.

As for patronage, the ear of the Crown was key; in a turbulent, troubled century, only those who were 'made of pliant willow, rather than stubborn oak' could hope to survive and prosper. The Gresham dynasty, Richard and his son Thomas for example, served with distinction variously as Lord Mayor, successful merchants or financial advisers to monarchs of markedly different religious persuasions (Geoff Pavitt, Gresham College lecture, 16 May 2009). In this regard, the name of the magnificent new bourse or Exchange, (Saunders, 1997) built with the perception and patronage of Thomas Gresham, is of real significance. It was instrumental in transferring the commercial and financial centre of Europe from Antwerp to London (Broodbank, I, 1921, 52). This was as a place for London's merchants to do business, with a shopping arcade to show off the finest products of its trading expertise. It was so beloved by Queen Elizabeth that, on her first visit on 23 January 1570, she insisted it be proclaimed the 'Royal' Exchange 'from henceforth, and not otherwise' (Wheatley, 1956, 173). And so it was.

Expanding worlds, expanding minds

Education was an essential ingredient of the expanding port that required a keen understanding of science and of navigation and a well-educated bureaucracy supported by literate and numerate officials. The literacy levels in London expanded dramatically in the 16th century, fuelled by the proliferation of published texts in a town famed for its printing presses. The city had a number of schools, such as Christ's Hospital, St Anthony's in Threadneedle Street, Merchant Taylor's in Suffolk Lane and St Saviour's in Southwark, as well as a major law school, based around the Inns off Fleet Street, Inner Temple and Gray's Inn (Weinstein, 1994, 48–9). But it did not have a university on the lines of Oxford, Cambridge or St Andrews, or the thirty or so similar foundations already operating across mainland Europe, from Seville to Krakow, from Louvain to Salerno. It is of interest to note that Sir Thomas Gresham, the financier, merchant and gun-founder, was well aware of that omission. In his last will and testament, he made provision for an eponymous multi-faculty college to be set up, at which seven professors would provide free lectures. The subjects included Astronomy, Geometry and Civil Law, as well as Divinity, Rhetoric, Physic and Music. Each lecture was presented twice, once in the morning in Latin (the European lingua franca), but repeated in the afternoon in English, for those who had not benefitted from a classical education (Wheatley, 1956, 70). Gresham College has been providing free lectures for Londoners from June 1597 to the present, and our own research project has been able to avail itself of these good offices (although our lectures and conference were delivered solely in English).

Technological factors that facilitated the revolution in oceanic navigation in the 16th century included the detailed cartography that developed alongside the traditional descriptive rutters (e.g. Waters, 1967; Alsop, 1996), much expanding their functionality, accuracy and geographical extent. Then there was the use of globes, building on those developed by Johann Schoener from 1522 (Baldwin, 2001, 63), and better instrumentation – a trade in which London, once again, had a particular expertise (Turner, 2000). The enterprises and expeditions facilitated by such advances were seen as a patriotic duty to God and the Queen as much as profitable business. The inducements that drove those voyages, from an English perspective, were summed up by the Gresham Professor of Astronomy, John Brereton. In his tract *A brief and true relation of the Discoveries of the North West Part of Virginia* he pronounced that they were for:

a) The Glory of God by planting of religion among those infidels;
b) The increase of the force of the Christians;
c) The possibility of enlarging the dominions of the Queen's most excellent majesty, and consequently of her honour, revenues and of her power by this enterprise (Baldwin, 2001, 97–8).

Richard Hakluyt (1522–1616), a man 'learned in and devoted to matters of geography' is best known today for his great opus *The Principal Navigations, Voyages…and Discoveries of the English Nation*, first published in 1589, and again in an enlarged three-volume edition between 1598 and 1600. However, he was also instrumental in the publication (as author, translator or facilitator) of many other books or pamphlets that underpinned or influenced much of the contemporary

Spreading the Word

Jacques Cartier, 1580, *A Short and Brief Narration of the two Navigations and Discoveries to the Northwest Parts called New France*, London (translated at Hakluyt's instigation and expense)
Richard Hakluyt, 1582, *Divers Voyages touching the Discovery of America and the Islands adjacent unto the same, first of all by our Englishmen and afterwards by the Frenchmen and Britons*, London
Richard Hakluyt, 1584, *A Particular Discourse concerning the great necessity and manifold commodities that are likely to grow to this Realm of England by the Western Discoveries lately attempted* (unpublished, but copy presented to Queen Elizabeth; document also known as the *Discourse on Western Planting*, written to promote Walter Raleigh's colonial interests
René Gouliane de Laudonnière, 1587, *A Notable History containing four voyages made by certain French captains unto Florida, wherein the great riches and fruitfulness of the country with the manners of the people hitherto concealed are brought to light*, London (translated from the French by Hakluyt)
Juan Gonzalez de Mendoza, 1588, *The History of the Great and Mighty Kingdom of China and the situation thereof, together with the great riches, huge cities, political government and rare inventions in the same*, London (translated from Spanish by R. Parke at the request of Hakluyt)
Thomas Harriot, 1588, *A Brief and True Report of the New Found Land of Virginia Discovered by the English Colony there seated by Sir Richard Grenville at the special charge and direction of Sir Walter Raleigh*, London (translated with Hakluyt's encouragement for German, French and Latin editions in 1590)
Richard Hakluyt, 1589, *The Principal Navigations, Voyages and Discoveries of the English Nation, made by sea or over land, to the most remote and farthest distant quarters of the earth at any time with the compass of these 1500* years, London (dedicated to Sir Francis Walsingham; most copies include an additional six pages recounting the 'Famous Voyage of Sir Frances Drake', his circumnavigation of the globe in 1577–80)
Albertus Meierus, 1589, *Certain Brief and special Instructions for Gentlemen, Merchants, Students, Soldiers, Mariners etc employed in service abroad or any way occasioned to converse in the kingdoms and governments of foreign princes*, London (translated by P Jones, with Hakluyt's encouragement)
Duarte Lopes, 1597, *A report of the Kingdom of Congo, a region of Africa and of the Countries that border about the same drawn out of the writings of Odoardo Lopez, a Portingall, by Philippo* Pigafetta, London (translated from the Italian by A Hartwell, encouraged by Hakluyt)
Richard Hakluyt, 1598–1600, *The Principal Navigations, Voyages, Traffics and Discoveries of the English Nation, made by sea or over land, to the most remote and farthest distant quarters of the earth at any time with the compass of these 1500 years. Divided into three several volumes, according to the positions of the regions whereunto they were directed*, London (vol. I dedicated to Lord Admiral Charles Howard; vols II and III to Sir Robert Cecil)
Jan Hugyen van Linschoten, 1598, *John Huighen van Linschoten. his discourse of Voyages unto the East and West Indies*, London (translated from Dutch by W. Phillip, encouraged by Hakluyt)
Johannes Leo Africanus, 1600, *A Geographical History of Africa, written in Arabic and Italian by John Leo, a Moor born in Granada and brought up in Barbary*, London (translated by J. Pory, encouraged by Hakluyt)
Antonio Galvao, 1601, *The Discoveries of the New World from their first original unto 1555. Briefly written in the Portugall tongue by Antony Galvano, Governor of Terante, the chief island of the Malucos: corrected, quoted and now published in English by Richard Hakluyt, sometimes student of Christchurch in Oxford*, London.
Jacob Corneliszoon van Neck, 1601, *The Journal, or Daily Register, containing a true manifestation and historical declaration of the voyage accomplished by eight ships of Amsterdam, under the conduct of Admiral Jacob Corneliszen Neck, which sailed from Amsterdam 1st March 1598 (to the East Indies). Showing the course they kept and what other notable matters happened unto them in the said* voyage, London (translated by William Walker, encouraged by Hakluyt, dedicated to Sir Thomas Smith, Governor of the East India Company)
Gerrit de Veer, 1609, *The True and Perfect Description of Three Voyages done one after the other by the ships of Holland and Zeeland on the North side of Norway, Muscovy and Tartaria, towards the kingdoms of Cathay and China, showing the discovery of the Straits of Weigates, Nova Zembla and the country lying under 80 degrees which is thought to be Greenland*, London (translated from Dutch by W. Phillip, encouraged by Hakluyt; dedicated to Sir Thomas Smith)
Peter Martyr Anghierra, 1612, *De Novo Orbe, or the History of the West Indies, containing the Acts and Adventures of the Spaniards which have conquered and peopled those countries, enriched with variety of pleasant relation of the manners, ceremonies, laws, governments and wars of the Indians. Comprised in eight decades*, London (translated from Spanish by R. Eden, encouraged by Hakluyt, dedicated to Sir J. Caesar, Chancellor of the Exchequer)

Table 2.3: Spreading the Word: works on exploration and navigation that Richard Hakluyt wrote or translated, or the publication of which he encouraged or facilitated (Payne, 2008, 88–114)

thinking concerning such voyaging. His impact and readership has been recently reviewed by Anthony Payne and it is from his important study that the catalogue of titles shown in Table 2.3 has been drawn (Payne, 2008, 88–114). It provides a literary context for the period in which the Gresham Ship, her owners, masters and crew were operative. It also shows one of the new ways in which information on new worlds was widely disseminated through the novel medium of the printed book, many of which were published in London.

London and the Gun Trade (Phil Magrath)

Voyaging on the high seas in the 16th century was a dangerous business. Quite apart from storms, treacherous tides and inhospitable coasts, there was the ever-present risk of piracy in an age when cargoes were valuable and neighbouring nations bellicose. Merchantmen travelled in convoy when they could and most were armed. Consequently a major port like London would have a ready market for guns, not just for the state's army and naval needs, but also for its merchant marine. This summary review of the gun trade in 16th century London is based on a lecture presented at Tudor Port of London conference in May 2009, supported by Gresham College. The principal sources for information on the foundries came from studies by Kennard (1986) and Ffoulkes (1969), while the research on gun-founding in the Weald was by Teesdale (1991), supported by the author's own research with the Exchequer Books at the Public Records Office.

The first serviceable cast-iron gun was made in 1543. Prior to that, guns were made in wrought iron or cast in bronze. We know of three foundries located in London, and the oldest of these was at the Tower where the Office of Ordnance had its administrative buildings and storehouses. The first guns cast here in 1354 were bronze bombards, but the industry was still too poorly developed by Henry VIII's reign to provide sufficient artillery for his needs and thus most had to be imported for his military campaigns.

It is not known where the Tower foundry was, but presumably it would have been close to the river, not only to facilitate the loading and unloading the necessary fuel and raw materials, but also to ease the transportation of the cumbersome finished products. It was, however, outside the actual confines of the Tower itself, well clear of the intra-mural gunpowder stores. A Thames-side location close by Iron Gate Wharf in St Katherine's seems most likely. Guns are depicted lying on Tower Wharf on both the 1544 Wyngaerde panorama and the Hayward and Gascoyne plan of 1597, and the latter also shows piles of shot near the White Tower. Our knowledge of the gun-founders operating at the Tower foundry during Elizabeth's reign is poor, with only one certain name, that of Henry Pitt in 1581, recorded as the Royal Gun-founder, Freeman of the Armouries Company and Warden of the Tower.

The second gun-foundry was at Houndsditch, near Aldgate, and had been producing guns since 1511. A map of 1572 marks the site with the legend 'The Gun-founders House'. It was here that skilled immigrant gun-founders worked, initially encouraged to settle in the area by Henry VIII to ensure that more guns could be made here rather than be imported. Although our knowledge of the practising gun-founders is scanty, we know of the Italian Arcano family and the Frenchman Peter Baude. Then there is the Owens family, whose involvement included armaments for the *Mary Rose*. One of their bronze guns has been recovered and bears the proud and patriotic inscription 'Robert and John Owyn brethren borne in the City of London the sonnes of an Inglish(man) made thys bastard (i.e. a non-standard weapon) AD 1537' (Rule, 1982, 165). This product of the Tudor port can now be seen on display in Portsmouth, together with the impressive remains of the ship itself. But detailed information concerning the gun-founders at Houndsditch is scanty. One account suggests that the foundry closed in 1550, but Samuel Owen, another Royal gun-founder, was in business here at the sign of the 'Guilden Goun' before his death in 1603.

The third known London foundry was located on the site of Salisbury Place to the south of Fleet Street where Salisbury Square stands today. It seems to have closed in 1545 following an association with the Arcancos.

By the time of Elizabeth's accession in 1558 the situation with regard to gun-founding had dramatically changed. Fifteen years had elapsed since the manufacture of the first cast-iron gun at Buxted in East Sussex in 1543 and now four other foundries were casting in the Weald. Here of course they were able to benefit from quality raw materials such as iron ore, timber for charcoal and fuel and river water to power machinery. However, the Government showed no particular enthusiasm for this new weapon for several reasons. Firstly bronze guns were regarded as more aesthetically pleasing than cast-iron guns. Secondly there was no particular shortage of bronze guns. Thirdly bronze weathered better, especially at sea. Fourthly they were safer as cast-iron was more brittle than bronze and if it failed when firing it failed catastrophically by exploding. Bronze tended to bulge and crack first before exploding, providing a vital warning. Fifthly Wealden gun-founders had not yet overcome the practical difficulties of casting heavier barrels. However, cast-iron guns had important advantages as they were lighter and cheaper to make than bronze guns. Additionally it soon became apparent that they were excellent in terms of ballistics and this was to have a major bearing on who wished to buy them and in terms of the market place. Importantly, nobody else was casting guns in iron in Europe during this period so the demand was always likely to be high.

London, of course, was the principal location for sales. William Camden described the shipping there in 1586 as 'a very wood of trees disbranched to make glades and let in light, so shaded is it with masts and sails.' In fact, as far as overseas trade was concerned, it was responsible for four fifths at a time when every other port nationwide – with the exception of Bristol – was in an advanced state of decay. The three bronze London gun-foundries were well placed to bring any products to the market place, but the situation was a little different for the foundries casting in iron. Due to their geographical location, in and around the Weald, it was a difficult and expensive business getting heavy goods to London (Teasdale, 1991).

Guns could be transported to London overland via East Grinstead, Godstone and Croydon. Alternatively, there were two waterborne options. Firstly, a northern route by cart from the foundry to a suitable place on the River Medway and then to Mill Hall near Aylesford via Maidstone. Here, guns were transferred onto hoys that would take them either to Tower wharf or onwards to an eastern coast port. It is unclear how or where the deal between gun-founder or his agent and merchant would have been struck, but Mill Hall was an important focal point. Secondly, a southern route from the foundry, by cart, to a suitable place on the River Ouse and thence to the port of Meeching (Newhaven) via Lewes. Once here, guns could be shipped around the Kent coast to London or westwards to English southern ports. More worryingly for the Government, some slipped across the channel to the continent.

There were plenty of potential buyers as the reputation for English cast-iron guns grew. The Spaniards, for example, were extremely short of artillery. Their usual supplies of bronze guns from France and Italy had ceased and they were certainly operating agents in the Low Countries with the specific aim of acquiring English cast-iron guns. Equally keen, but of less concern to the Elizabethan government, were Dutch Calvinists who had since 1569 been fighting the Spaniards at sea as privateers. Additionally the high seas were by all accounts infested with pirates of many different nationalities, including English, all desirous not only to arm their vessels with these guns and to continue to menace trade, but even perhaps to sell them on to Spaniards.

The Government was clearly worried by this and in 1568 came the first in a series of measures to control the export of cast-iron guns. One of the foremost Wealden gun-founders and 'Royal Gonnestone Maker', Ralph Hogge, complained to the Privy Council that his trade was being adversely affected by fellow gun-founders exporting in noticeable quantities. Here was the dilemma for the authorities. Men like Hogge had to be encouraged to stay in business, but the opportunities for sales in the domestic market were limited. Even by the end of Elizabeth's reign bronze guns were preferred to arm the navy, yet this important industry needed to be maintained and controlled. Three Wealden foundries could most likely have met the national demand for cast-iron guns but with ten foundries active in 1570, there would inevitably be a glut. The only way round this was to permit limited exports and Hogge was accordingly granted the exclusive right to manufacture guns and shot for export. In 1574, a similar license was granted to Sir Thomas Gresham, who established the major Mayfield foundry in 1569. Hogge was complaining again in 1573 but by now ten foundries were productive, He wrote that 'yor enemie is better fourneshed with them then or owne country ships ar'. He cited sales to Holland, Flanders and France in particular and suggested closer supervision of coastal ports such as Hastings, Portsmouth, Rye, Plymouth and Bristol where large consignments were apparently evading the export licence.

The following year the Privy Council was convinced enough to tighten the laws governing manufacture and restrict the number of guns cast 'for the only use of the Realm'. The order was issued from Hampton Court and on 3 February 1574 Richard Pedley, one of the Royal Messengers, was sent to the Weald to order the attendance of all iron-masters at the Tower. Of the 77 people identified, 58 attended and so seriously was the matter taken that each was bound under a bond of £2000 not to manufacture any cast-iron ordnance without the Queen's special licence. All iron guns cast had to be delivered to Tower Wharf where they could be sold only to English merchants or to foreigners with rights of residence who owned English ships or were in partnership with English ship owners. Further, any ship owners making purchases had to account for their ordnance within one month of a return from their voyage. The buyers were to enter into a signed undertaking which was deposited with the Ordnance Office that they were purchasing guns solely for the defence of their own vessels and not with the intention of selling them abroad. Finally the gun-founder himself had to be able to provide full details of any sales he had made.

Sadly the only bonds that have survived are a series of 148 from 1590, although they provide a fascinating glimpse into the trade of cast-iron guns in London at this time. So, for example, we find recorded the names of purchasers and their occupations – mostly ship owners and/or merchants, with the name of the ship and its home port. Also noted is the merchant selling the guns, the type of gun and the numbers involved. There seems to have been a group of merchants regularly dealing in guns, for example, John Chapman, Lactancius Fox, Richard Marshall and Giles Fleming. The origins of the vessels are of interest too; for example, three of the ships were from the Low Country ports of Enkhuizen and Flushing. All the guns featured in these bonds are of the smaller type and represent guns destined, in theory, to arm predominantly English merchantmen working from English coastal ports. The falcon, for example, fired a 3-pound shot, the minion 4-pound shot, the saker a 5–6-pound shot and the demi-culverin a 9-pound shot.

A certain Mr Orsell is a notable presence in these accounts, acting as an agent for the Low Countries, while there is only one occasion when an actual gun-founder – John Phillips – acts as the vendor. Phillips is an interesting character whose activities in the market place at this time would have caused concern to the conscientious Ordnance Office official. His career was a riches to rags story, starting promisingly at either the Tower foundry or at Houndsditch making bronze guns before becoming the Queen's Royal gun-founder. He then moved to the Tonbridge area perhaps lured by the more lucrative prospect of casting guns in iron and his name is associated with the export of guns from

Rochester in 1580, 1585 and 1587. But in 1581 he became involved with a notorious gun-trafficker called Elliot, was eventually convicted and in 1588 was serving time in Marshalsea prison. The Royal Armouries collection at Fort Nelson has a fine example of one of his son's bronze sakers. This was cast by Richard Phillips probably at Houndsditch at the sign of the Fireball in 1601.

London's pre-eminent position in the market place can be shown from evidence in the surviving Port Books. These were essentially the records of the imports and exports into a port for the collection of revenue, Although many have been lost and those surviving are often damaged and illegible one is able to compare tonnages of cast-iron guns being shipped either to the continent, to London or to other English ports through Rochester and Meeching between 1579 and 1599.

Once again, Teasdale (1991) has provided the figures. Through Meeching 1057 tons of guns were shipped to all locations and of that amount 719 tons (68%) went to the continent, 178 tons to London (17%) and 160 tons (15%) to other English ports. Through Rochester, 907 tons was shipped to all locations and of that – by far the lion's share – 854 tons went to London (94%) and a mere 52 tons (6%) to other English ports; so there were no apparent exports from Rochester to the continent in this period. Collectively then, 1964 tons of guns were recorded by the Port authorities as leaving both ports in this period. 719 tons or 37% went abroad, 1033 tons or 52% ended up in London, and 214 tons or 11% went to other English ports. Twenty years of trading is not an insignificant period and supports the importance of London in the trade of this commodity. It would appear that the 1574 requirement to bring guns to the Tower of London had been relaxed. What these figures do not show of course were the inevitable clandestine or unrecorded exports.

That cast-iron guns were illegally exported, whether via London or through other routes, is not in doubt for several reasons. First even from London, guns could have been shipped onwards illegally, for example, through the notoriously corrupt Office of Ordinance itself. Secondly if Ralph Hogge's accusations were accurate, merchants like John Harman of Lewes were more than prepared to take risks for high financial gain. He was cited as having illegally sold the guns of the Wealden gun-founders Alexander Fermour and Nicholas Fowle abroad. Thirdly two other gun-founders, Davy Willard and Michael Weston, for unknown reasons were excused from signing the 1574 bond. This would have given them free reign to sell to whomever they pleased. Fourthly there is no doubt that some of the Spanish Armada's guns were English and of cast iron. Indeed, nearly 30% of that fleet's guns were iron, but it is unknown what the ratios were of cast-iron to wrought-iron. All cast-iron guns would have been English because nobody else in Europe was making them. Fifthly so worried was the Privy Council that Sir Francis Walsingham, Elizabeth's principal secretary, spymaster general and Privy Councillor employed a certain William Pistor with the grand title of 'searcher for illegally exported ordnance'. Teasdale believes that between 1568 and 1600, when annual production varied between 500 and 1100 tons a year, between a half and three-quarters of this tonnage found its way across the Channel, legitimately or otherwise. Since all ships required ballast, one probable way to avoid prying eyes was to stow clandestine shipments below.

Chapter 3: Conservation Studies

by Dean Sully and Kelly Domoney

The Finds Study Programme

The Finds Study Programme for the Gresham Ship Project was designed to carry out and disseminate technical, analytical and methodological research into the hull, cargo and associated remains of the site investigated in the Princes Channel, Thames Estuary, by Wessex Archaeology in 2003 and 2004 (Firth, 2006; Auer and Firth, 2007; see also Volume I of this monograph). A significant element of the Finds Study Programme has been the interventive conservation of that finds assemblage. Maritime archaeological materials are vulnerable to unpredictable change following excavation and consequently their post-excavation study and long-term survival relies on such interventive conservation procedures (Cronyn, 1990).

Following excavation and recovery, the finds assemblage was initially stored with Wessex Archaeology (Fig. 3.1). When University College London began its research programme in April 2008, the assemblage was relocated to temporary storage at Fort Cumberland with English Heritage (EH), where work was carried out as part of the Gresham Ship Project Conservation Internship from October 2008 to April 2009 (Fig. 3.1). Many of the objects were then moved to the laboratories at the Institute of Archaeology in University College London (UCL), forming a focus for course work related to the MSc Conservation for Archaeology and Museums programme (Domoney, 2009). Following conservation stabilization, the finds archive will be transferred to the care of the Southend Museums Service (see Appendix for finds list). A selection of those finds will ultimately be displayed in a purpose-built gallery focusing on Thames wrecks in a brand new museum building overlooking the estuary where the Gresham Ship was recovered. Although this will be some years in the future, a temporary exhibition that featured some of the Gresham Ship's artefacts has already been presented in the Central Museum at Southend from June to October 2012.

The significant costs of maritime archaeological excavation and post-excavation processes have important implications for the ability of developers and commercial archaeological units to negotiate effective outcomes for marine archaeological assemblages. The role of UCL within the Gresham Ship Project was not to act as a commercial archaeological unit and conservation service provider. Rather, the finds assemblage has been used as a research resource for graduate students at UCL's Institute of Archaeology as part of MA and MSc course work (Table 3.1), utilised in targeted dissertation research, practical projects, work placements and internships.

Reliance on student work for the completion of the finds research projects can be problematic, since it depends on the availability and interest of individual students in this specialist area of archaeology. Nevertheless, in this particular case, access to the finds assemblage provided a key opportunity for interdisciplinary research with the resource used for graduate research and teaching for the Conservation, Museum Studies and the Technology and Analysis of Archaeological Materials graduate programmes.

The advantage of this project taking place at a university, such as UCL, is the opportunity to develop a broad based academic approach to the challenges presented by maritime archaeology from diverse specialist perspectives. This provides the potential to focus on broad themes, such as the methodological studies of existing archaeological processes, as well as enabling highly focused research questions to be developed on specific elements of the wreck assemblage (Gould, 1983; Lenihan, 1983; Murphy, 1983; Watson, 1983; Martin, 2000, Bowens, 2009). There is a need to build capacity in the expertise of maritime archaeology in the UK. The use of graduate students on specialist archaeological courses at UCL is seen to be one very important way in which specialist expertise can be developed to meet future challenges. For conservation students, the project

Figure 3.1 Temporary storage of Gresham Ship finds assemblage: on the left at Wessex Archaeology in 2006; on the right with English Heritage in 2008 at Fort Cumberland, Casement 23 (photos Gresham Ship Project).

Edward Salter, 2006, *Getting into Deep Water? The Princes Channel and the preservation of Historic Shipwreck Sites*, Unpublished UCL MA Managing Archaeological Sites Dissertation, Supervisor Tim Williams.
Kelly Hanson, 2007, *Analysis of leather finds from the Gresham ship assemblage*, Unpublished UCL MA Principles of Conservation Dissertation, Supervisor Dean Sully.
Katie Smith, 2007, *A conservation strategy for three cast iron cannons from the Gresham Ship*, Unpublished UCL MA Principles of Conservation Dissertation, Supervisor Dean Sully.
Kelly Domoney, 2007, *The application of condition monitoring techniques to the Gresham ship Hull timbers in Horsea Lake*, Unpublished UCL MSc Conservation for Archaeology and Museums Dissertation, Supervisor Dean Sully.
Kelly Domoney, 2008/9, Six-month Gresham Ship Project Conservation internship, based at Fort Cumberland, English Heritage. MSc Conservation for Archaeology and Museums, Supervisor Dean Sully.
Domoney, K., 2009, *Gresham Ship, Princes Channel, Thames Estuary: Conservation of an Elizabethan Shipwreck Assemblage*, English Heritage Research Department report series, **102**. Swindon: English Heritage. Publication illustrates how the Gresham Ship Project trained future researchers in maritime conservation.
Thomas Birch, 2009, *The Gresham Ship: An investigation in the iron bars. Analytical project to investigate the provenance for the iron, lead and tin ingots, metal small finds*, Unpublished UCL MSc in the Technology and Analysis of Archaeological Materials Dissertation, Supervisors Thilo Rehren and Marcos Martinón-Torres.
Gresham Ship Exhibition Project 2009: 'A Thames Odyssey The *Gresham Ship* journey so far.....'. Production of museum display to coincide with public outreach programmes. MA Museum Studies, Museum Communication Practice, Supervisor Theano Moussouri.

Table 3.1 List of UCL student projects associated with the Gresham Ship Project

Gresham Ship Assemblage of Artefacts	**Archive Location**
Four guns	Fort Nelson, Royal Armouries
Large ship timbers, iron anchor, gun carriage, ship ballast, iron bars	National Diving Centre, Stoney Cove
101 Small finds	Southend Museums

Table 3.2 Location of Gresham Ship finds (See Appendix for complete list of objects)

provided a practical and theoretical understanding of the conservation of marine objects. Confidence was gained through practical experience of preventative measures (for example, wet storage, care while performing interventive treatments) and decision-making about investigative cleaning methods (for example, mechanical or chemical?), desalination processes, drying and stabilization techniques.

The Finds Assemblage

The finds assemblage revealed in the process of excavation and investigative conservation consisted of four iron guns, 42 iron bars and 101 small finds including three lead and five tin ingots, six leather objects, six small wooden finds, eight metal object fragments, copper and tin alloy small finds (salt holders, spoons, a bowl, a jug), plus bricks, stone ballast, rope, glass and ceramic fragments (Table 3.2). The partial and vestigial nature of the assemblage (a common feature of rescue archaeology) limits the potential of the subsequent research. However, there were in addition some 25 ferrous concretions recovered from the wreck site, some or all of which may have accumulated around and then obscured significant finds; the role of the objects initially hidden within the concretions was therefore of particular importance in this Finds Study Programme, offering an opportunity to add significantly to our understanding of the ship and its crew (Auer and Firth, 2007, 14). With this in mind, rather than considering the assemblage as a diminished representation of a complete shipwreck, our approach has been to look at the finds assemblage as a series of snap shots connecting us with events in the past of the ship, as links in a chain that connect the present with the past.

The documentation and conservation of the four guns has been undertaken at the Royal Armouries at Fort Nelson, Portsmouth reported in Chapter 4 of Volume I of this monograph, pages ?–?). The remainder of the assemblage – including the concretions, artefacts and the metal cargo – form the focus of the Finds Study Programme. Details of the metal cargo are discussed in Chapter 4, while the remainder of this chapter focuses on investigative conservation of the concretion assemblage and the conservation of the small finds.

Temporary Storage

Upon recovery marine archaeological objects are exposed to a change in environmental conditions, which can cause rapid deterioration if not controlled (Robinson, 1998; Watkinson and Neal, 1998). The favourable conditions of low oxygen availability, low light and low temperature that may have given rise to the long term survival of archaeological materials are likely to give way to drier conditions of increased oxygen, light and temperatures (Pearson, 1987; Oxley, 1998; Yeager, 2005). The transition from excavation to post-excavation storage needs to be carefully managed to limit adverse changes in the recovered archaeological material. Post-excavation storage conditions should not adversely affect subsequent examination and conservation treatments and should be relatively inexpensive and easy to maintain (Jones, 2003, 350). Following their recovery in 2004 by Wessex Archaeology and prior to conservation treatment, finds were held in temporary wet storage, bagged and labelled in eight plastic crates immersed in tap water in a large scaffold tank. Even where marine finds are kept wet following excavation (see Figure 3.1), the increased availability of oxygen, combined with available water and the presence of chlorides, leads to secondary corrosion in the corrosion products of metal objects (Smith, 1991, 762). For organic objects, the risks of biological deterioration are increased with increased availability of light, heat and oxygen (Cronyn, 1990, 271). A beneficial side effect of immersed storage in fresh water is the potential for the diffusion of chlorides from the corrosion products into the water. This has extended the desalination period for all finds stored immersed in tap water since excavation. The conductivity of the tap water was monitored and storage water periodically refreshed. Alkaline inhibitive or oxidizing solutions are often used with ferrous metal objects to prevent secondary corrosion, but this is not appropriate for assemblages of mixed materials (Watkinson and Al-Zahra, 2008). The use of low temperatures and the exclusion of light is preferred to the use of biocides, when managing temporary storage of waterlogged organic materials (Kirsten, *et al.*, 2012).

Conservation and maritime objects

Archaeological conservation aims to ensure that artefacts are able to function within the requirements of the archaeological process. Initially this may involve the provision of advice during project start up and initiation (Satchell and van Rensburg, 2006). Such advice is essential to anticipate the scale of potential conservation work and develop cost-effective strategies prior to excavation taking place, but is often absent in rescue archaeological projects. During field work conservation intervention may be required to assist with the delicate excavation of vulnerable objects: this is particularly complex when excavating marine sites. Consequently conservation intervention frequently starts with the temporary storage of assemblages prior to the research and long term potential of excavated material being assessed. Investigative conservation, as part of excavation and post-excavation assessment, provides essential data for the interpretation of the archaeological resource (Cronyn, 1990). This, combined with interventive conservation treatments, aims to render the conserved object assemblage suitable for handling, study and eventual storage and display in standard museum and archive conditions, by improving its physical strength and resistance to damage from environmental factors (Cameron, *et al.*, 2006). This has been the focus of conservation work within the Finds Study Programme.

The physical stability of the object assemblage and the retention of contained information that can be revealed through archaeological investigation remains the primary focus of the archaeological conservation process. Increasingly however, greater importance is given to the cultural significance of objects (Muñoz Viñas, 2005). The process is therefore seen to result in the selection of cultural values in the conserved object; as a consequence, certain values will be retained, maintained or enhanced, whilst others will be diminished, altered or removed. The cultural significance of the conservation object is now understood to provide the focus of a conservation process that is revealed by investigative procedures (Avrami, *et al.*, 2000; Pye, 2001).

Conservators operating within a values-based conservation process are, therefore, unable to focus simply on the physical materiality of the objects in their care (Clavir, 2009, 145; Sully, 2014). Accordingly, the aim of conservation is to enhance the preferred cultural values of material heritage, while maximizing the potential of people in the future to access their own preferred values (Muñoz Viñas, 2009, 56).

Within the conservation process, it is recognized that there may be many 'truths' that can be revealed in the 'true nature' of a conserved object (Sully, 2014). Conservation intervention can be seen to crystallize certain interpretations of these truths in the physical fabric of the object, whilst removing evidence of others (Muñoz Viñas, 2005). The conserved object is necessarily an edited version of all the interactions that link the current state of the object with its past states. It presents a selection of certain truths, materialized as a conserved object, in a clear narrative distilled from the mass of individual stories that complicate an understanding of the past. Its complexity is reduced in order to render the object as a specifically constructed representation of itself. Hence the conserved object reveals only one authored version of the past, rather than having any claim to reality other than that revealed in the process of conservation (Sully, 2014).

Caple describes conservator's decision-making process as compromise between the three goals of conservation: revelation, investigation and preservation (2000). The goal of 'revelation' may sometimes conflict with the goal of 'investigation', for example, one may seek to remove concretions from a heavily corroded iron artefact in order to discover information about it, but this prevents the conservator from presenting the artefact as found, as a corroded iron artefact. Likewise the conservator may wish to present the object as a corroded iron artefact, but that may not allow a stable object to be produced or allow sufficient investigative procedures to reveal contained information about the object. Archaeological objects from maritime shipwrecks provide a stark example of how the conservator's decisions affect objects. Waterlogged organic objects excavated from a maritime archaeological site, for example, are often significantly altered in form from the original manufactured object and are vulnerable to rapid deterioration in physical condition once excavated and as part of the archaeological process. A conservation response is therefore required to stabilize these objects between the excavation environment and post-excavation environments. These actions are always limited by available time and resources. However, maritime archaeological artefacts will require active conservation intervention in order to be useful within an archaeological process.

Authentic objects

Artefacts from shipwrecks have complicated life histories from their production, ownership, exchange, use, adaptation, loss, burial, excavation and post-excavation processing. In conserving these objects a conservator is faced with decisions about which elements of these biographical events to preferentially preserve in the physical fabric of the conserved object. If we consider the two extremes of a spectrum of change in the nature of the object, the conservator might to seek to present the conserved object as a representation of its original manufactured form or alternatively as an archaeological artefact as found within the shipwreck. These possibilities may be limited by the physical condition of the object, but are also a reflection of the expectations of the conservation process within the specific circumstances of the conservation moment. To preserve an object in its original form, as an act of restoration, would mean reversing the changes that have occurred since the object was manufactured. This would involve the removal of evidence of the object's use, the impact of the sinking of the vessel, the time spent submerged and any effect of the archaeological recovery process. This would require drying the object, removing the effects any changes that have occurred to the object over time (such as later additions, change of form, deformations) and removing all marine deposits within the object (such as soluble salts), around the object (such as insoluble concretions) and the physical associations between objects linked together by changes caused by the burial environment. By removing the evidence of the shipwreck, a significant portion of the artefact's history is removed from its physical fabric. This change dramatically alters the way the object is understood and deemed significant. As an alternative, the aim for conservation could be to conserve 'as found', as a representation of a shipwrecked artefact. This would suggest that the object should be kept wet, with any marine deposits retained. This might appear attractive in restricting the expense of conservation interventions; however, long term storage and display of waterlogged artefacts is technically difficult and, in the majority of cases, objects are only considered to function within an archaeological archive once dried and stabilized (Ganiaris and Starling, 1996; Goodman and Barnard, 2005; Björdal, *et al.*, 2007; Institute for Archaeologists, 2009; Museum of London Archaeological Archive, 2013).

Drying the object with its marine deposits unaltered is, in most cases, unlikely to produce an object in a stable condition and may have a disastrous effect on the structural integrity of the artefact. The preservation of a 'shipwreck artefact' therefore is more likely to stand for conserving an object using a 'minimal intervention' approach to altering the object as found. Likewise, preserving the artefact 'in its original' form simply stands for the notion of removing any evidence of notable layers of an artefact's history from its physical fabric. In reality, there is a gradient of intervention that will seek to balance the production of a 'plausible' conservation object from the constraints manifested by the object undergoing treatment and the time and resources available to the conservation process (Sully, 2014).

A conservation process that seeks to redefine the object as a representation of its original form is a selective choice that diminishes the object as palimpsest. The desire to render the object in a form other than it currently appears – to reflect a specific expectation about how the object should look (either at point of manufacture prior to use, prior to burial/disuse, prior to discovery or prior to collection); to conform to a specific intellectual process (such as art history, archaeology or anthropology); to reflect a current perception of a well presented archaeological museum object – needs to be resisted. Any such change needs to be effectively justified within a decision making process that defines a clear aim for the conservation treatment that considers the implications for the current and future object (Sully, 2014).

The desire to retain the potential for the conserved objects to reflect the links between the present and the past as shipwreck objects has been the guiding principle of the conservation of the Gresham Ship finds assemblage. The actual conservation undertaken represents a pragmatic response to this guiding principle. This approach is clearly demonstrated in conserving the concretions, in which conservation is a creative process, the product of which is a series of unique conservation objects. The approach of a sculptor or fossil hunter is required in revealing a plausible object from an amorphous concretion. The resulting form of the object is a product of the choices and skills of the conservator in revealing something of value within the altered layers of the concretion. In the absence of objects, their form can be recreated, as the voids are materialized with

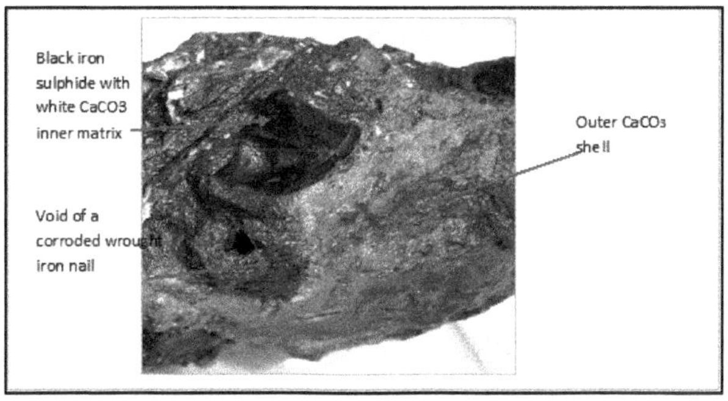

Figure 3.2 Inner black iron sulphide with $CaCO_3$ inclusions and outer $CaCO_3$ shell (photo Gresham Ship Project)

casting resins. The conserved object becomes a hybrid of modern polymer, residual object materials and the creative possibilities provided by the conservation process.

Conserving concretions

A significant component of the excavated finds assemblage were the 25 ferrous concretions containing objects in various states of preservation. The formation of marine concretions around ferrous objects can be seen to seen to hinder the identification of the archaeological object inside and obscure the extent of its corrosion and deterioration. Alternatively. However, it can be seen to provide the raw material for a conservation process to reveal a unique and plausible shipwreck object. Conservation was therefore undertaken to investigate the significance of the concretions, to reveal primary archaeological information, to reveal plausible objects and to conserve identified finds.

Concretion formation

Concretions form on iron artefacts buried in marine sediment (anoxic conditions) and those exposed to seawater on the seabed (aerobic conditions). As ferrous (Fe^{2+}) and ferric (Fe^{3+}) ions diffuse outwards they are replaced by chloride (Cl^-) ions from the seawater; therefore, there will be a concentration of Cl^- ions in the iron corrosion of the object (Turgoose, 1989). Iron corrosion products interact with materials in the marine environment in close proximity to the corroding iron object. As iron is non-toxic to marine organisms, iron artefacts are rapidly colonized resulting in a build-up of marine growth (North and Macleod, 1987, 77). Primary marine growth consists of calcium carbonate ($CaCO_3$) exoskeletons of calcium-depositing organisms such as corals and tube-worms (Florian, 1987, 12). Marine growth creates a hard, rough surface layer on the iron, intermixed with sediments, debris, other wreck material and marine organisms. This layer can retard the corrosion of iron by providing a low-porous barrier between the metal and seawater. Corrosion at the surface of the iron does still occur inside the concretion; the trapped seawater, subject to depleted oxygen levels, will generate an acidic solution (pH 4.8) (North and Macleod, 1987, 77). The solution slowly dissolves the calcite matrix and replaces it with a black iron sulphide matrix.

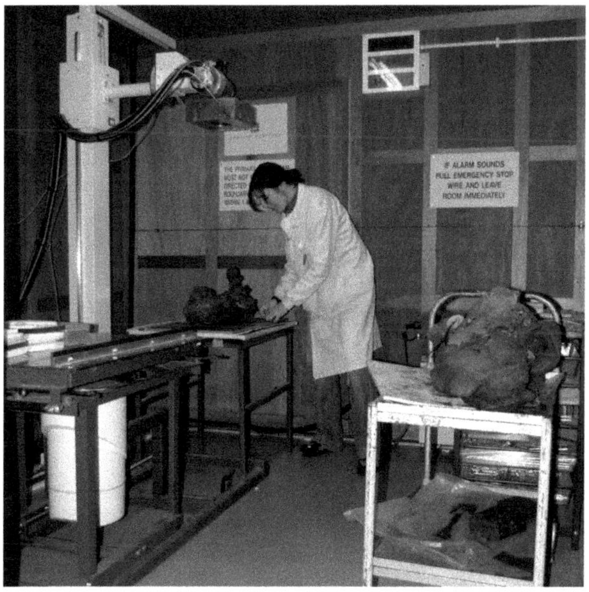

Figure 3.3 X-radiography of concretion assemblage at English Heritage, Fort Cumberland, in 2008 (photo Gresham Ship Project)

The acidic-iron solution passes through the corrosion product region into pores and crevices in the $CaCO_3$ concretion (Scott and Eggert, 2009, 123). The $CaCO_3$ present is gradually replaced with a hard 'iron cementing matrix' of Fe^{2+} and Fe^{3+} ions as it slowly dissolves in the acidic solution (North and MacLeod, 1987, 77–8). As a result of these reactions, the inner concretion consists of a replica of marine growth formed from iron corrosion products and an outer shell consisting of precipitated $CaCO_3$ (Fig. 3. 2).

Condition assessment

Upon initial examination of the concretions, no information was available concerning the nature of potential finds contained within; therefore, appropriate conservation materials and methods for processing the assemblage were investigated. A pilot project was initiated to determine the potential of the concretions to reveal archaeological and condition information via X-radiography (Fig. 3.3). From this initial assessment, a sample group of concretions was processed in order to develop suitable treatment methods, prior to conserving the remainder of the assemblage.

Documentation and investigation utilised photography, X-radiography and illustration (Figs 3.4 to 3.10). X-radiography was conducted using an AGO HS 225kV Hi-Stability X-radiograph system with Kodak MX125 and Kodak AK film. X-radiography of the assemblage indicated the concretions contained a variety of artefacts in differing states of preservation; well preserved (type 1), completely corroded (type 2) and partially corroded with core intact (type 3). A fourth category of concretion was also identified, in which the contents were of indeterminate form and preservation (type 4). This was largely due to the density of the larger concretions, in which features within the X-radiograph image were often

Type 1	Well-preserved metal object
Type 2	Void of completely corroded iron objects
Type 3	Partially corroded iron object with intact solid core
Type 4	Indeterminate form and preservation

Table 3.3 Concretion Types 1-4, describing state of preservation of metal artefacts within concretions

difficult to identify (Table 3.3; Plates 1 and 2 (Figs 3.4 to 3.8) and Fig 3.9).

Concretion Treatment

The approach to the treatment of the concretions initially involved mechanically removing loose deposits of sand and sediment from the outer surface. Where possible X-radiographs were used as a guide to the 3-dimensional form of the contained objects and the depth of concretion (Lang, 2005).

Digital photography of X-radiographs proved to be a useful tool in identifying slight differences in density, reducing the need for further repeat radiographs. Figure 3.9 presents two digital photographic exposures of an X-radiograph of concretion containing partially corroded hook (GSP99). A short exposure in the upper image reveals the overall shape of the concretion. The longer time exposure in the lower image helps to reveal the shape and corroded edge the hook tang (left hand side of image). For Concretion GSP86 X-radiography indicated non-ferrous metal-alloy artefacts to be well-preserved within the concretions, associated with part-mineralized wood and voids of corroded wrought iron bar and nails (Plate 2, Fig. 3.7). The shape and surface details of completely corroded metals were preserved in the surrounding concretion and characterized by dark voids in the X-radiographs. In general, bulbous voluminous areas of dense concretion surrounded the voids. Tracings of X-radiographs on Melinex (polyester sheeting) were found to be useful in recording associated positions of artefacts and voids within concretions. Tracings and X-radiographs were used to guide the excavation procedure (Fig. 3.10).

Several techniques for processing the concretions were evaluated for suitability, which were adapted to the nature of the concretion, the state of preservation and location of any contained objects. Options included stratigraphic excavation and/or fracturing the concretion into large sections to reveal contents, with the aim of retaining extant objects and casting void cavities to reveal the form of lost objects.

The following processing methods were applied to concretions containing objects in different states of preservation.

Type 1 Concretions: concretions containing well-preserved metal objects (Fig. 3.4)

Given the small number of discrete finds present in the Gresham ship assemblage, a key aim of initial processing was to remove the bulk of the concretion in order to

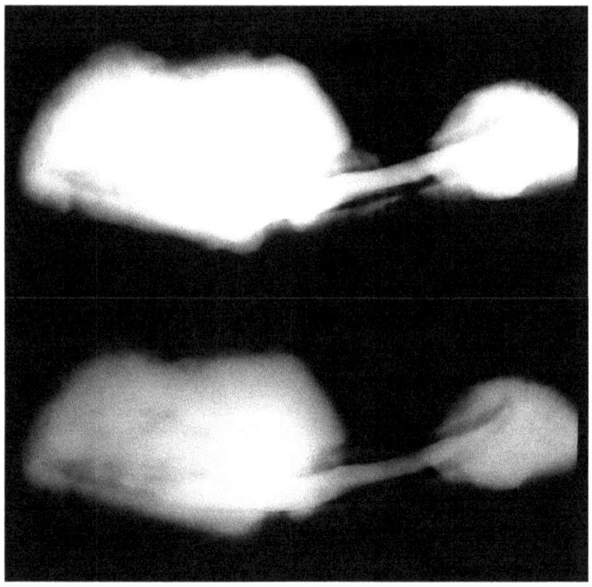

Figure 3.9 Two digital photographic exposures of an X-radiograph of Concretion GSP99 containing a partially corroded hook (photos Gresham Ship Project)

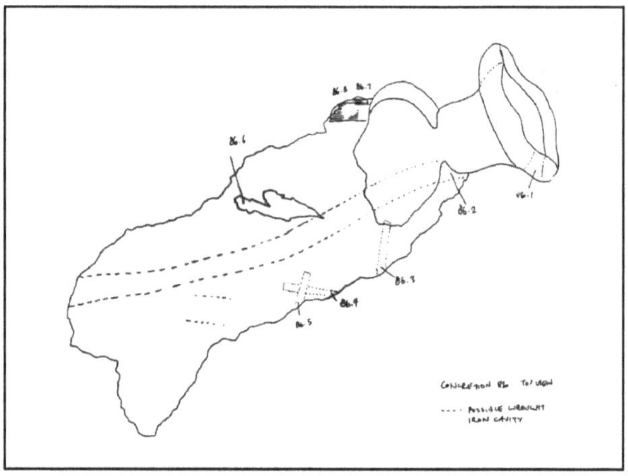

Figure 3.10 Tracing of X-radiograph used to determine associated positions of artefacts and voids within concretion GSP86 prior to excavation. See Plate 2, Fig. 3.7

identify any objects and materials that were present for subsequent treatment and stabilization for the finds archive (Figs 3.4, 3.5 and 3.6 on Plate 1). Well-preserved metal objects were characterized by presence of metal with limited corrosion and weak bonds between the object and the concretion (North, 1987, 210).

Treatment options to reduce the concretions included the use of circular saws, sandblasting, air abrasive and dental drills (Redknap, 1984, 133; Carpenter, 1990). Strong acids have been used to dissolve concretions from cast iron artefacts, but these can be slow, ineffective and potentially damaging to the corrosion products and metal core (North, 1987, 211). Cryogenic freezing with liquid nitrogen was used successfully to crack and remove the concretions from cannon from the shipwreck *Trial* (MacLeod, 1987, 54).

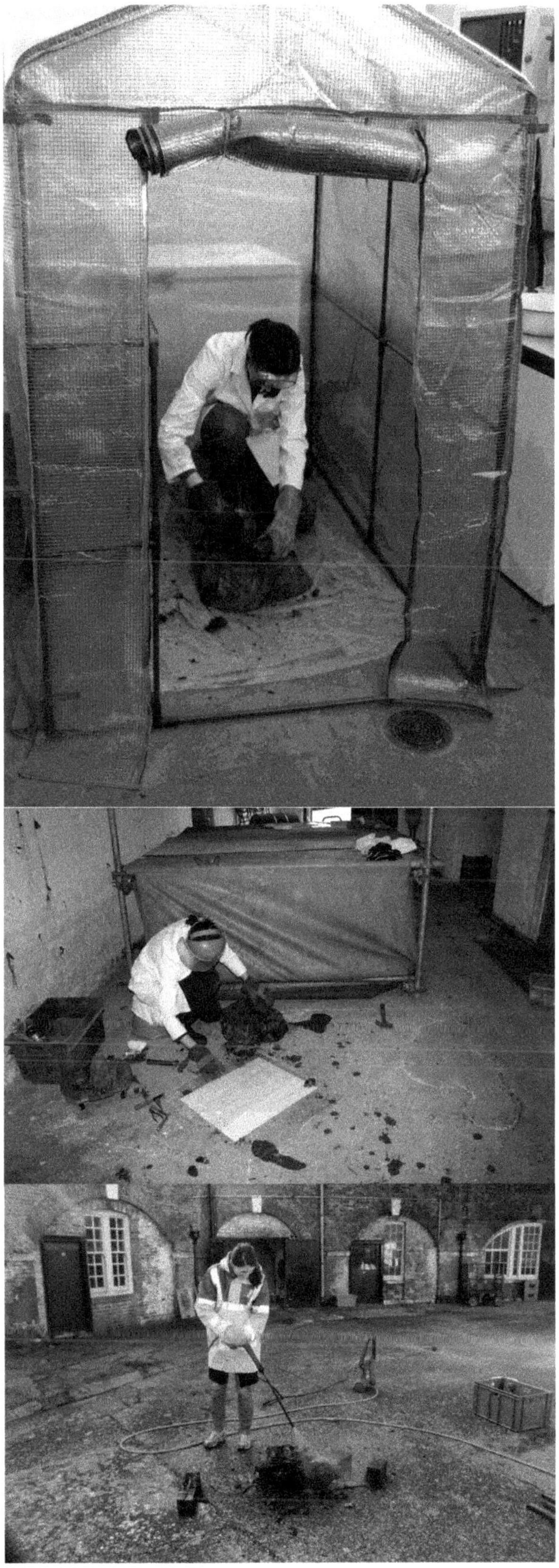

Figure 3.11 Processing concretions at the English Heritage facility in Fort Cumberland (2008/9): (above) inside plastic greenhouse containment area; (centre) inside storage Casement 23 and (below) outside storage Casement 23 (photos Gresham Ship Project

The most effective method employed on concretions which did not appear to contain large voids, ceramics or fragile organic objects was to strike the concretion with a large flat hammer, in order to cleanly break the concretion into a series of large sections to reveal contents. A series of strikes were made perpendicular to the surface of the concretion, which loosened small sections of concretion, which were then removed by hand. The size of the tools used depended on the thickness of the concretion. For small concretions or those with a thickness of 5 cm or less, small hammers or geopicks were found to be adequate for processing. For larger concretions, or those with a thickness of 5 cm or more, large flat hammers and stone chisels proved most effective. Concretions were struck in areas of weakness, such as cracks, or at the intersection of different materials, for example, where stones or shells lay in the concreted $CaCO_3$ surface. This process was difficult to control, especially where dense corrosion encapsulated softer less dense contents, leading to unpredictable fracturing of the concretions. Concretions were kept wet with a water spray during processing to avoid uncontrolled drying (Fig. 3.11). Once the bulk of the concretion had been removed, smaller tools such as vibrotools, dentist picks and small chisels were employed to reduce the mass of the concretion in order to separate objects. Where required, final traces of concretion were removed chemically or mechanically in the laboratory depending on material type and condition.

Type 2 Concretions: concretions containing voids of completely corroded iron objects (Fig. 3.5)

The aim of processing concretions containing completely corroded iron objects was to cast the remaining voids in order to preserve the form and surface details of the original object (Muncher, 1988; Mardikian and David, 1996; Arnold and McAllister, 1998). The corroded iron within the voids was usually in the form of a black liquid. X-ray fluorescence analysis of the residue indicated iron sulphide, based on the presence of high iron and sulphur peaks in the X-ray spectrum. The voids were cleaned of residue before a casting material could be applied. To do this concretions were cracked into two or three large fragments by striking with a hammer and chisel at a perpendicular angle along a pre-determined inscribed line. The aim of this method was to crack directly through the void and create segments with clean break edges in order to allow for tight reassembly. The iron sulphide residue was then removed by washing with small brushes, wooden tools and pipe-cleaners in running water. Where necessary, two access points were drilled into the void; one for pouring in the casting agent, the other for allowing air to escape. One hole was drilled in concretions with voids that extended through to the surface. The interior surfaces of the voids were partially dried by rinsing in acetone to enable the casting agent to cure effectively. Moisture remaining in the concretion acted as a natural release agent once the casts were fully cured. Concretion segments were reassembled using 3M self-adherent flexible veterinary wrap (Plate 2, Fig. 3.12, upper centre). Visible cracks or gaps were plugged with Plasticine (chalk, mixed oils, colouring), Cling Film (low

Casting material	Curing time	Viscosity	Working time	Tensile strength	Colour	Curing method
Silastic 3483 silicon rubber base and Silastic 83 curing agent	24 hours	17000 mPa.s	90-120 minutes	3.5 MPa after 7 days	White	Condensation reaction
Tiranti Polyester Resin (with 4% Butanox M-50 Liquid Hardener) (Methyl Ethyl Ketone Peroxide, 33% in Dimethyl Phthalate) mixed with Tiranti Iron Weighting Filler in 60%/40% v/v ratio	20 minutes	180-600 cps	10 mins		Brown	Exothermic reaction
Araldite 2020 epoxy resin with powder pigments	40–50 minutes	150 mPa.s	30 mins		Brown	

Table 3.4 Casting properties of epoxy, polyester resin, and silicone rubber casting materials

density polyethylene) and Plastazote (closed-cell cross linked polyethylene foam).

Casting materials

Three types of casting materials were found to be useful for casting voids of different sizes and complexity (Larsen, 1981). Silicone rubber was useful to replicate complicated forms due to its flexibility on removal. Epoxy resin was effective in casting small intricate voids in a one-piece mould, due to its low viscosity and low shrinkage. Pigmented epoxy casts had a good appearance, but were light in weight and liable to fracture during excavation of the concretion. Polyester resin with iron weighting filler proved useful for large several-part casts; due to its short curing time and low cost (Table 3.4). Iron weighting filler provided additional strength to the cast, preventing damage during excavation and providing an appropriate colour for the resulting cast. However, increased shrinkage occurred during curing and the iron filler presents a potential problem of corrosion post treatment. Due to the strengthening properties and short curing time, polyester resin with iron weighting filler was less likely to be damaged in the process. This was selected as the standard material for casting concretion voids.

Fully cured casts were removed, using gentle hammering, followed by vibrotools in the same manner previously described for concretions containing well-preserved metal objects.

Casting the Void: Concretion GSP25

Concretion GSP25 provides an example of the process of revealing an object cast from a void cavity. The X-radiograph provided clear guidance for the expected location and shape of the concretion cavity (Plate 1, Fig. 3.5, centre). Following initial cleaning and investigation, a chisel and hammer were used to remove superficial encrustation and to crack open the concretion (Plate 2, Fig. 3.12, top image). The crack occurred perpendicular to the void creating two segments with clean break edges, which enable for tight reassembly of the fragments for casting. The void cavity was cleaned and rinsed with acetone. Tiranti General Purpose polyester resin with iron weighting powder (see Table 3.4 for mixing ratio) was poured into the opening cavity in the concretion fragments (Plate 2, Fig. 3.12 upper centre), once the cured, the cast sections were mechanically removed from the concretion using a hammer and chisel (Plate 2, Fig. 3.12 lower centre and bottom). An epoxy resin (Araldite 2020, bisphenol A-epichlorohydrin) was used to join the sections of the cast pieces together to reveal a nail or a spike fastening for deck planks (McCarthy, 2005). The cast accurately records the change of form from a square cross-section into a round cross-section. The bottom tip of the nail appears to be broken, the cast had captured this detail well and this suggests the possibility that this tip of the nail could have been hooked (Fig. 3.5 on Plate 1 bottom image).

Type 3 Concretions: concretion containing void of partially corroded iron object with intact solid core (Figs 3.6 and 3.13 on Plates 1 and 3)

X-radiography of GSP99 indicated the concretion to contain a hook-shaped cavity, potentially with a solid metal core (Plate 1, Fig. 3.6). In order to determine whether the core was solid or consisted of fragmentary or liquid iron corrosion products, the concretion was broken into four sections in the same manner as previously described for objects with complete voids. The concretion was found to contain a solid wrought iron core surrounded by liquid iron corrosion. The void was cleaned, sections reassembled and cast in three stages with Tiranti General Purpose polyester resin with iron weighting powder. The three-part cast was necessary as the central core was liable to block the flow of the casting resin. Hook and surrounding cast were desalinated and air dried followed by reconstruction with Paraloid B-72 (polymethyl acrylate copolymer) from a tube (approximately 85% concentration). The visible iron core appears to have been stabilized by the consolidating effects of the polyester resin cast. The object as a whole is solid and stable.

Concretions GSP203

The technical difficulty of revealing the contents of concretions can be seen in the treatment of concretion GSP 203. Before treatment, the object consisted of a large crust of $CaCO_3$ and iron corrosion products that obscured the form of the object(s) or any voids as shown in Fig. 3.14 (Plate 3) An X-radiograph appeared to show a narrow void running through the centre of the object, the ends contain denser material and there may therefore have been metal remaining in these areas

The concretion was fractured using a large hammer perpendicular to its surface, revealing a cylindrical iron object running through the centre of the bulk of the concretion. Although there was a significant amount of iron remaining, there were also voids suggesting that this surviving iron would not be strong enough to survive any attempts to remove it from the concretion without some form of consolidation. The interior surfaces of the voids were dried with acetone prior to filling with Tiranti General Purpose polyester resin with iron weighting powder. This consolidated the remaining iron, creating a combination of a cast and part of the original object. As the casting was revealed, it became clear that only the core was made up of resin; the surface was still very friable and liable to loss. The surface of the fragments was wrapped with a temporary facing protected with Nylon Gossamer and impregnated with 10% (w/v) Paraloid B-72 (ethyl methacrylate copolymer) in acetone. Due to the extremely fragile nature of the iron surface, some of this was lost during processing; however, the majority was retained and consolidated. This provides support for the fragments during desalination immersion in de-ionized water with 1% v/v triethanolamine as a corrosion inhibitor. The fragments of the complete bolt are shown in Fig. 3.15.

The length of this bolt as shown in Fig. 3.15 is approximately 310 mm with the head around 55 mm across. The diameter appears to slightly taper towards the bottom which is expected of bolts to make them easier to insert, it ranges from 20 mm at the bottom to 27 mm just below the head. This is probably a clinch bolt, the end of this bolt was left projecting about a half a diameter above the timbers to be fastened and then a ring was placed on top. The head of the bolt was then hammered in order to spread out the head of the bolt further and tighten it over the ring thereby securing the timbers in place (McCarthy, 2005, 69–71).

Type 4 Concretions: concretion containing unidentified objects (Plate 2, Fig. 3.8; Plate 4, 3.16)

X-radiographs of the majority of the Gresham Ship concretions did not reveal readily identifiable features or objects. Concretion GSP202 was chosen as a test case to assess processing options which could then be applied to similar concretions (Plate 4, Fig. 3.16). The concretion was gently cracked open using a flat hammer and stone chisels. This was carefully undertaken to avoid uncontrolled fragmentation of the outer concretion (and therefore destroy any internal voids) or damage any contained objects.

During initial processing, the concretion was found to contain the form of an iron cannon powder chamber (GSP202.1 – Plate 4, Fig 3.17). At this point, the remainder of the concretion was excavated stratigraphically, using light hammering in order to fragment the hard iron sulphide and calcium carbonate concretion matrix. During excavation a hammerhead with partially mineralized wooden shaft (GSP202.2) and a fragment of rope (GSP202.3) were discovered and positions recorded on the Melinex plan.

Preliminary research indicated the powder chamber (dimensions 450 x 250 x 250 mm) to be similar in style to iron examples excavated from two Spanish Armada shipwrecks: *El Gran Grifón* and *La Trinidad Valencera* off Fair Isle, Scotland and Kinnagoe Bay, Northern Ireland dating to 1588 (Martin and Parker, 1988, 222 nos 16 and 21).

Not all concretions were as straightforward as concretion GSP202. In many cases, the contents of the concretions were difficult to interpret with evidence of lost metal elements represented by a combination of different cavities and wood from which specific features could not be identified. Concretion GSP20, for example, revealed discrete finds, such as a silver spoon, ceramic fragments, wrought iron chain fragments and cast voids, but other fragile elements were more difficult to preserve. Such features were photographed and where possible a mould was created with either silicone rubber and/or cast with polyester resin.

After excavation recovered artefacts were conserved by material type (see below). The remainder of the removed concretion was processed into gravel-sized pieces of approximately 10 mm in diameter using a flat hammer, in case small objects, such as coins, were present. The rubble was discarded, with a few samples retained as examples of the results of concretion processing.

Removal of chlorides

The most destructive feature of submersion of iron objects in seawater is the presence of chloride salts, which concentrate within objects and their corrosion products (Reguer, *et al.*, 2007). Removal of chlorides is crucial in mitigating the long-term adverse effects of the corrosion process. Upon excavation from their concretion, iron objects (for example, the powder chamber GSP202.1 and chain links GSP90.1) were immediately immersed in a sealed tap water environment to retard further deterioration, prior to desalination treatments being initiated.

A variety of desalination techniques have been developed for use on marine iron artefacts (North and Owens, 1981; Rees-Jones, 1972; Smith, 1991; Gonzalez, *et al.*, 2007, Wang, *et al.* 2008). Methods vary in intensity, cost, level of efficiency, equipment and

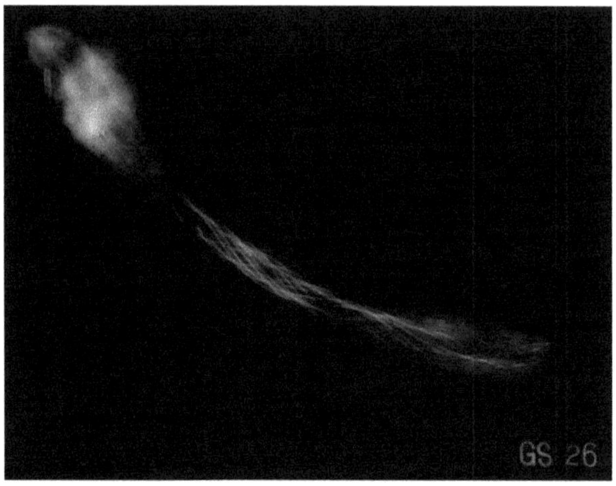

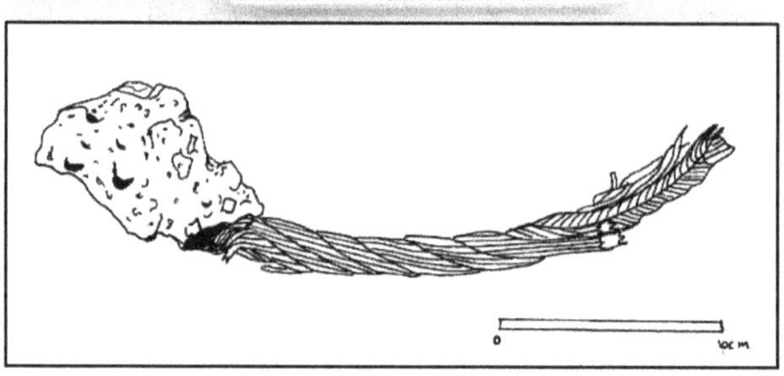

Figure 3.18 GSP Concretion no 26: dramatic X-radiograph revealing remains of modern metal cabling (above); illustration of object (right) before and after partial concretion removal (photos Gresham Ship Project)

resources. The most commonly used techniques are water diffusion, alkaline sulphite reduction and electrolysis. Electrolysis is the preferred choice of many conservators when the equipment, expertise and resources are available (Hamilton, 2010). It provides an efficient and effective method of removing chlorides, loosening concretions and reducing iron corrosion products to the denser magnetite (North, 1987, 216). A less efficient, but more straightforward method of chloride removal is water diffusion, with the object being soaked in successive baths of tap water with an added corrosion inhibitor. Washing in 2% w/v sodium hydroxide (NaOH) in a container at least five times the volume of the object is recommended by North (1987, 221), measured using a chloride ion selective electrode (Wang, et *al.* 2008). This is then followed by immersion in distilled water with a corrosion inhibitor, such as sodium nitrate ($NaNO_2$) at 1000 ppm (North, 1987, 222–3).The iron objects recovered from concretions were transferred to a 5% w/v solution of sodium sesquicarbonate ($Na2CO_3$/ $NaHCO_3$) in tap water, which stabilizes the iron by neutralizing the acidity, preventing oxygen infiltration and forming a passivating film on the surface of the iron. Ferrous metal objects have been desalinated or are undergoing desalination, using this method, monitored through pH measurement to ensure that a level between pH 11–12 is maintained for periods of over one year (North and Macleod, 1987; Carpenter and MacLeod, 1993; Weizhen and Chunchun, 2005, 101). Measurement of pH was with Merck Alkalit® pH7, 5–14, BDH indicator strips 0–14, Merck Acilit® indicator strips 0–6 and Merck Neutralit® indicator strips 5.0–10.0. Chlorides were measured with chloride strip test, Merckoquant® (Merck) 1.10079 Chloride Test (Cl^-) and silver nitrate test (Odegaard, *et al.*, 2000).

Other alkaline corrosion inhibitors were used: for example, 1% v/v triethanolamine as a corrosion inhibitor was used with composite materials, such as wood/iron composites (for example on hammer GSP202.2) and polyester resin/iron composites (for example on metal bolt GSP203). Non-ferrous metals, other inorganic and organic objects were desalinated in tap water without the addition of corrosion inhibitors, monitored with conductivity measurements (using an InoLab pH/Cond 750 meter and waterproof meter DiST by Hanna (HI98311). Water baths were monitored and changed weekly/monthly until no increase in conductivity was measured over one month/three months. A two-stage desalination using tap water followed by de-ionized water was used in a specific cases (for example, non-metal inorganics) and carefully monitored in the case of copper alloy/concretion GSP201.1. De-ionized water immersion can have detrimental effects on some materials, such as tin alloys. Tin alloy will be adversely affected by immersion in solutions below pH 8; therefore de-ionized water and often tap water will need to be buffered prior to use (Hamilton, 2010).

Summary of Concretions finds

From the 25 concretions processed, a range of discrete objects and ship-related artefacts were revealed, along with the retention or partial retention of the concretions themselves. Objects or casts of objects that were excavated from a concretion were allocated sequential part numbers, based on the finds number of the concretion (for example, GSP86.1 to 86.7). A selection of the small finds will be discussed further in this chapter. However, details of each object and their conservation treatment can be found as an open access research data base. This provides a complete digital archive of object files containing conservation treatment records and associated analysis and before, during and after treatment images.

Metal and metal/organic composite objects were recovered from twelve concretions (GSP66, 86, 90, 93, 94, 95, 99 200, 201, 202, 203, 209). A combination of casting and object retrieval techniques were required to reveal partially preserved metal objects (for example, GSP66, 99 and 209). Object casts were recovered from concretion voids, as the sole record of the contained metal object for three concretions (GSP25, 83 and 96).

Five concretions (GSP26, 27, 216, 217 and 218) were prepared to present a tangible record of the stages in conservation, manifest as an object. Concretion GSP26 was retained as a partially processed concretion (desalinated and air-dried) (Fig. 3.18). Other concretions such as GSP27, were prepared as a partially processed concretion with cast elements in place, to present the process of concretion casting.

No information about contents was recovered from five concretions (GSP12, 29, 66, 70 and 97). These concretions were processed into small fragments to ensure no object information was present. The process in each case was documented and the concretion fragments were discarded.

Metal small finds (Tableware)

Conservation of tin-alloys

Three tin-alloy objects underwent active conservation; two salt holders (GSP14 and 86.1; Plate 5, Fig. 3.19), originally identified as candlesticks and a spoon (GSP86.7). As few examples of tin objects survive from marine burial environments, these objects were deemed significant for the interpretation of the site (Cronyn, 1990, 211; MacLeod and Wozniak, 1997, 118). All three objects had been selected for an exhibition on the Gresham Ship at University College London in May 2009 and were required to be conserved to display standard. The aim of the conservation process was to identify appropriate and effective conservation procedures to remove iron sulphide and calcium carbonate concretion from tin-alloys.

Analysis

Qualitative X-radiograph fluorescence (XRF) analysis of exposed metal areas indicated each object to be manufactured from tin with varying amounts of alloying metals (Fig. 3.20b).

Condition assessment

Visual examination and initial investigative cleaning of a salt holder (GSP86.1, dimensions: 108 x 77 x 33 mm) and spoon (GSP86.7, dimensions: 79 x 30 x 4 mm), both excavated from concretion GSP86, indicated the presence of strongly adhered iron sulphide concretion as well as calcium carbonate marine encrustation. As concretion covered approximately 80% of each surface, it was difficult to determine the condition of the underlying metal. However, exposed sections of the outer surfaces consisted of a fine grey protective patina typical of tin oxide corrosion (Cronyn, 1990, 211; MacLeod and Flecker, 2001). Calcium carbonate marine encrustation on the surface of a second salt holder (GSP14, dimensions 74 x 65 x 25 mm) measured between 2 mm and 7 mm in thickness. Investigative cleaning revealed a paste-like black and grey corrosion layer directly below the encrustation in which original surface features including the moulded design were present. The paste-like nature of the tin corrosion layer therefore affected removal methodology of the harder concretion crust.

Fig. 3.20a XRF analysis: using hand-held portable InnovX XRF instrument (photo Gresham Ship Project)

Treatment

The tin-alloy group was identified as a test case to document different approaches to their conservation. Mechanical removal was attempted on both salt holders. However, due to the strong adherence and density of the concretion compared to the underlying metal, this method was abandoned in favour of chemical means to reduce the risk of physical damage to the objects.

An experiment using different concentrations of dilute hydrochloric acid (HCl) in tap water was conducted on salt holder GSP86.1 (Table 3.5).

1. 1% v/v HCl localised application in cotton wool over 48 hours
2. 2% v/v HCl localised application in cotton wool over 48 hours
3. 5% v/v HCl localised application in cotton wool over 48 hours
4. 2% v/v HCl total immersion bath over 24 hours
5. 5% v/v HCl total immersion bath over 24 hours

Table 3.5 Chemical cleaning tests carried out on salt holder GSP 86.1

Immersion in 5% v/v HCl in tap water proved to be the most efficient method for removing the iron sulphide as the acid gradually dissolved calcium carbonate inclusions, which in turn loosened the sulphide matrix. This was chosen to remove encrustation adhered to salt holder GSP14.

Following concretion removal, both salt holders were desalinated to remove residual HCl. Desalination consisted of immersion in a running tap water bath for five hours followed by five static tap water baths over a two-week period. Conductivity and pH levels of the baths were regularly monitored. During desalination both objects exhibited active, localized corrosion in the form of white spots occurring on deteriorated sections of metal. White products formed after two days on GSP 14 and after seven days on GSP86.1. The formation of white

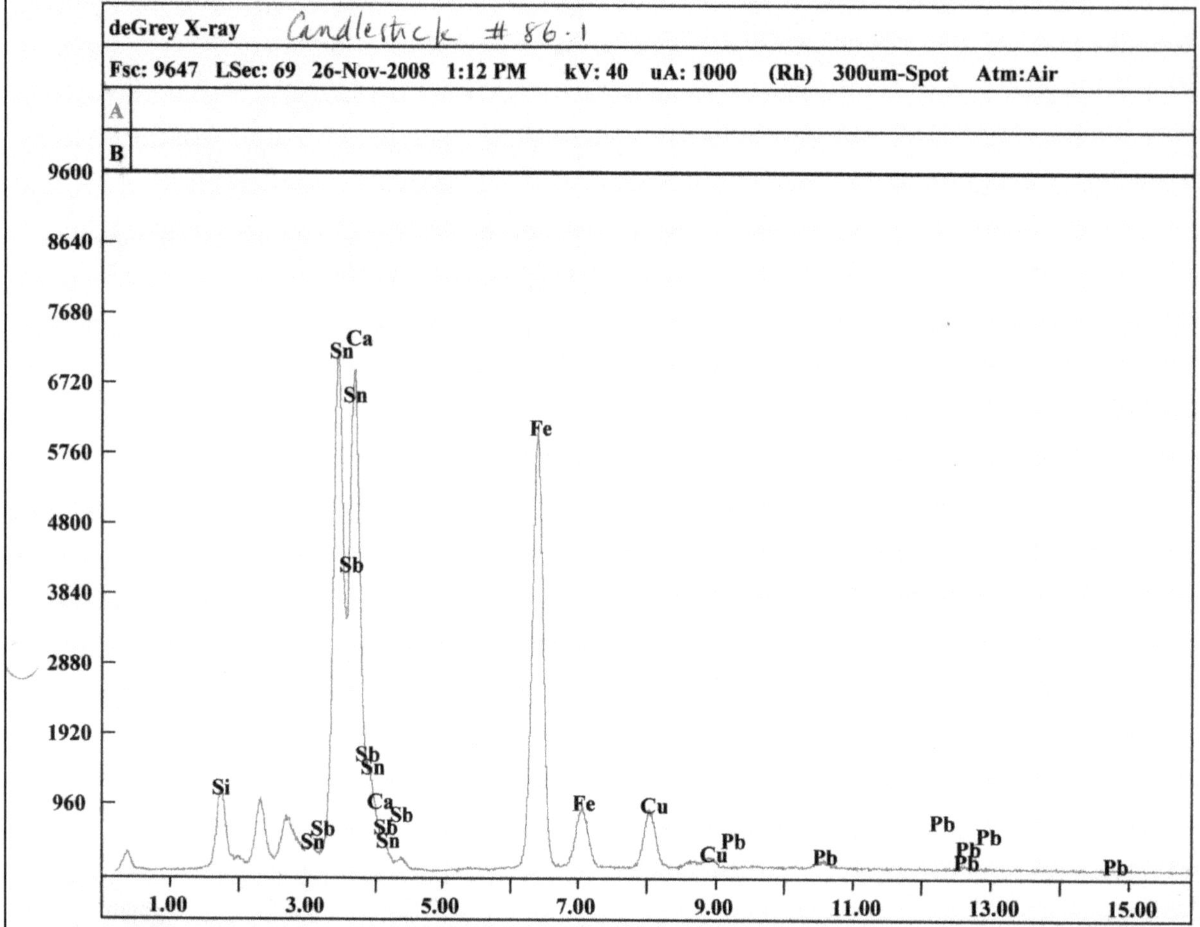

Quantitative Results

Element	Inte	Weight%	Atomic%
SiK	204.32	13.88	31.91
SnL	1365.79	59.86	32.56
SbL	6.85	0.38	0.20
CaK	680.09	12.72	20.50
FeK	1272.99	11.80	13.64
CuK	184.37	1.08	1.10
PbL	23.26	0.28	0.09

Stage X: 13.49 Y: 63.53 Z: 20.55

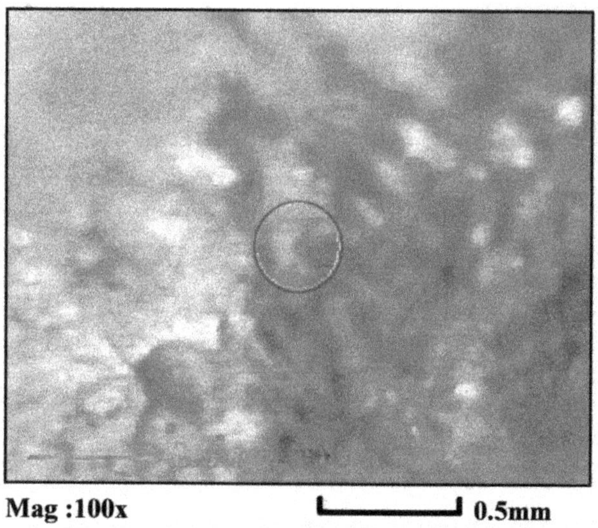

Figure 3.20b XRF analysis: pXRF data for surface analysis of salt holder GSP86.1

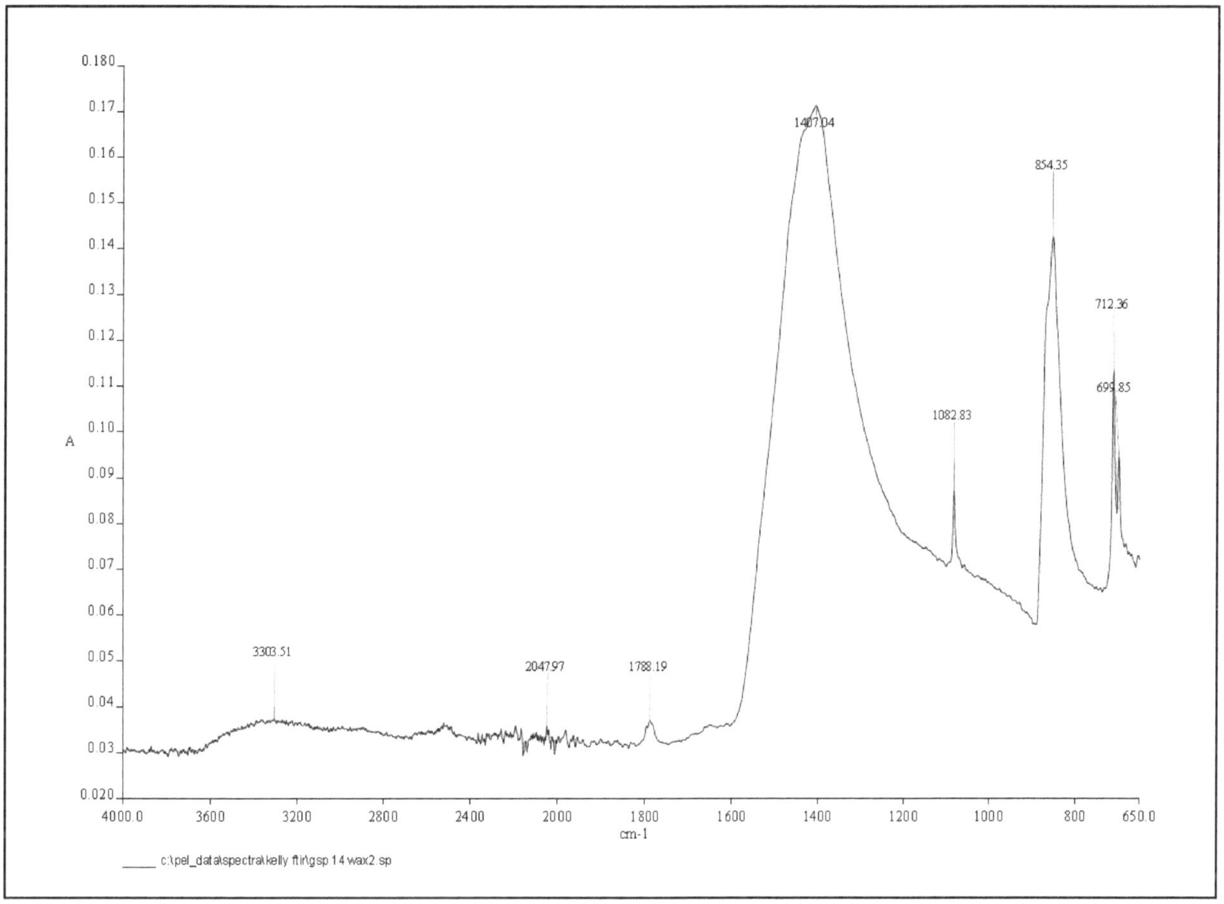

Figure 3.22 *Results of Fourier transform infrared analysis (FTIR) on green wax-like substance located inside salt holder GSP14*

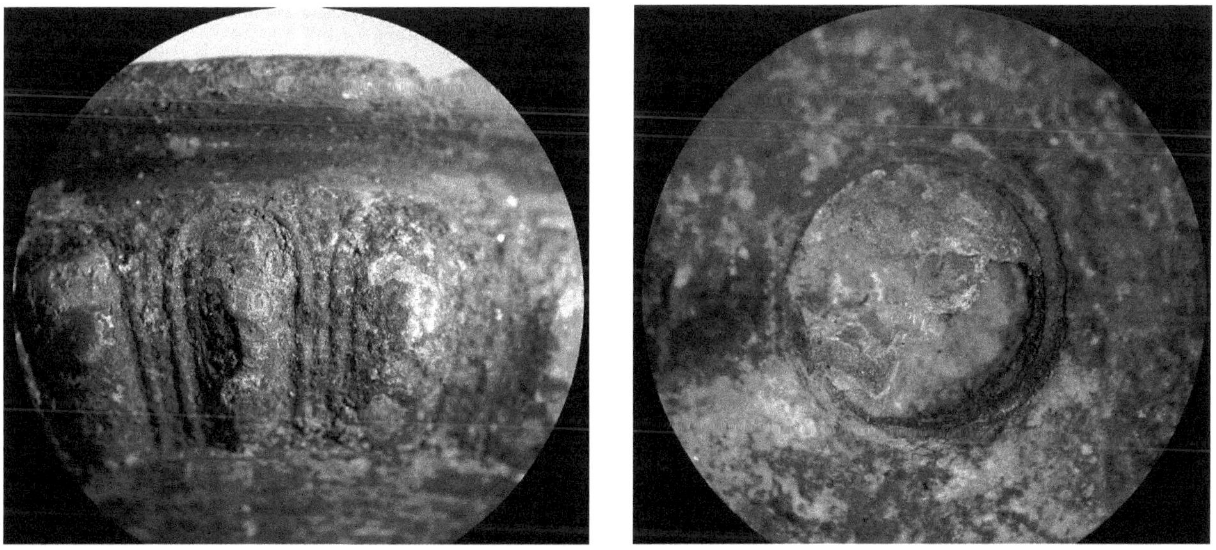

Figure 3.23 *Salt holder GSP14: left detail of moulding after HCl treatment; right wax-like substance in cavity of salt holder sampled for FTIR analysis (photos Gresham Ship Project)*

products is likely to be due to the pH of the water bath, as tin has the propensity to corrode in pH of less than 8 (Hamilton, 2010). Corrosion products may have formed faster on GSP14 than on GSP86.1, due to the advanced corrosion of the alloy prior to immersion. Each object was dewatered through an acetone bath for 24 hours, prior to air drying.

Chemical treatments proved extremely effective at producing objects with fine surface detail whilst leaving all corrosion layers intact (Figs 3.19 and 3.21 on Plate 5 and Fig. 3.23a). The pilot treatment emphasised the importance of maintaining high pH of immersion solutions for tin-alloy objects, even in areas with hard tap water.

Post-treatment analysis

After treatment, examination of one of the salt holders (GSP14) under optical microscopy indicated the presence of a green wax-like substance inside the cavity and in areas around the rim (Fig. 3.23 – White and Page, 1992). This has particular relevance given the initial identification of these objects as candlesticks. The substance was sampled and its composition analysed using Fourier transform infrared spectroscopy (FTIR; Fig. 3.22). The resulting spectrum was compared to reference samples of beeswax and beef tallow, common materials used in candle manufacture in the 16th century. No comparative bond-peaks were present. Further comparative FTIR analysis with lead and tin carbonate samples from the Infrared and Raman Users Group online spectral library (www.IRUG.org) resulted in similar peaks in the 700, 1100, 1450 and 1750 wavenumbers (cm-1) regions. Results suggest that the substance consists of a tin or lead corrosion product or residual marine encrustation.

A straight forward desalination and air drying treatment was carried out for the group of lead ingots (GSP16, 84, 85) and tin ingots (GSP33, 81, 91, 213). Lead and tin alloy metals are vulnerable to acidic conditions and are more stable when dried and protected from volatile organic compounds (Pearson, 1987, 113; Barker, 2003, 48). Therefore, this group of objects was desalinated in changes of tap water and once conductivity levels had stabilized at levels that matched the initial tap water readings, they were dried quickly to limit the amount of time spent immersed in water.

Spoons

Mechanical cleaning was chosen to remove iron sulphide concretion from the surface of a spoon that was revealed within concretion 86 (GSP86.7, dimensions 108 x 77 x 33 mm). Scalpel cleaning was effective in removing thin layers of concretion (1 mm in thickness), but was liable to result in scratch marks. Vibrotools were tested in dense areas of concretion. However, the vibrations were considered to be too damaging to the underlying object. Air abrasion without aluminium abrasive powder proved useful in removing dense areas of concretion. However, rapid removal resulted in a matted effect on the metal surface, possibly due to residual aluminium powder in the pressure chamber. Therefore, a combination of these techniques was carefully applied to limit their adverse effects.

Removal of the concreted layer revealed a stem with a square cross-section and a moulded decorative design along two sides ending in a volute scroll at the base of the bowl (Fig. 3.24). In a comparison with thirty-five Tudor and Stuart period pewter spoons excavated at riverside sites in Southwark, only one similar example is recorded and identified as a continental import (Egan, 2005; 112, fig. 102 no. 540). A maker's mark may have been present at the base of the spoon's bowl; however, due to heavy corrosion in this area, the remains of a mark are not visible or detectable by X-radiography.

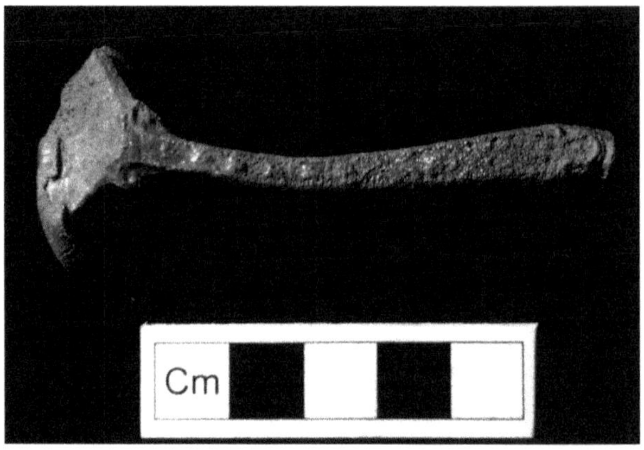

Figure 3.24 Spoon GSP86.7 after conservation (photo Gresham Ship Project)

The processing of concretion GSP20 revealed another spoon (GSP20.1, dimensions 152 x 83 x 14 mm), ceramic fragments, bone fragments, iron pin fragments and void casts (Plate 6, Fig. 3.25).

Mechanical cleaning using scalpels and air abrasive was used to reduce the outer layers of the hard concretion. To reduce the potential surface damage from mechanical removal on the soft silver surface, chemical cleaning was undertaken. A solution of 5% v/v dilute hydrochloric acid in de-ionized water was locally applied to the spoon's surface, so that the attached ceramic fragment was not affected. This revealed excellent surface detail, including a human figure at the end of the spoon handle, decoration along the handle on front and back and a circle at the top of the spoon bowl (inner surface) which may be a makers mark or hallmark. After cleaning in dilute hydrochloric acid residual chlorides were removed in four baths of running water each for 6 hours. The object was then desalinated in changes of tap water monitored weekly over three weeks using a Hanna EC/TDS Conductivity meter, prior to being air-dried.

Copper alloy jug GSP208

Copper alloy jug GSP208 is a rim fragment, broken across at the base of the neck with curled, sharp edges. This object was not associated with a concretion and when found exhibited a brown and green metal surface (Fig. 3.26) (MacLeod, 1982, Scott, 2002). XRF analysis suggests that the object is composed of brass (copper (Cu) 64–75%; zinc (Zn) 8–10%).

An attempted desalination using 5% w/v solution of sodium sesquicarbonate in tap water (Florian, 1987, 12) was abandoned after two days, when the bath water turned blue, indicating copper corrosion products were being washed out of the object (North, 1987, 235). In this solution, hydrolysis of the CuCl and $Cu^2(OH)3Cl$ occurs with the release of chloride ions. Due to the speed of this reaction, the object was removed from the alkali bath to passive storage in tap water. Desalination continued in 100% tap water for three weeks; conductivity levels were checked weekly using a Hanna EC/TDS Conductivity meter, after which the water replaced. This was repeated until the conductivity level matched that of the initial

Figure 3.26 *Copper alloy jug fragment GSP208: left before conservation; left after conservation (photos Gresham Ship Project)*

reading. The desalinated object was air dried in atmospheric conditions and regularly monitored for salt crystallization. The object was fully dry after two weeks, indicated by monitoring weight change to a stable level. The object was then packed for inclusion in the archive in a Stewart Box (polyethylene) with Plastazote (polyethylene foam) support and acid free tissue, with silica gel providing a desiccated environment to minimize further corrosion. The object will be assessed for stability prior to archiving, to determine whether any further stabilization is required, such as the addition of benzotriazole (BTA) corrosion inhibitor and protective coating (Paraloid B-44).

The Copper alloy/concretion GSP201.1

The concretion debris surrounding iron gun GSP201 was found to contain the remains of a copper alloy vessel and a ceramic jar. These were separated into discrete lumps of concretion during corrosion removal at Fort Nelson and transported to UCL Institute of Archaeology for conservation. The ceramic jar concretion GSP201.2 is discussed below. The copper alloy object concretion (GSP201.1) will be described here (Fig. 3.27).

The copper alloy object GSP201.1 was a large circular form (diameter 432 mm) with a relatively intact exterior surface. Portable XRF (pXRF) surface analysis indicated that the alloy was brass (58–70 % copper (Cu); 25–31% zinc (Zn), see Table 3.6. The primary aim of the conservation treatment was to reveal information about the manufacture and use of the object, to prepare the object for long-term deposition in an archive and for potential display in a future exhibition. The fine impression of the associated gun makes it a good candidate for display as a shipwreck object. Its value lies in its story-telling capability; it is visually striking and informative of the overall story of the Gresham ship. Its typological identification and manufacturing information potentially add to the body of evidence supporting the Tudor period date for this vessel.

The impression of the gun's surface can be seen on the interior concave surface of the concretion, leaving the distorted external side of the copper alloy object exposed (Fig. 3.27 and Plate 7, Fig. 3.29). The object had crumpled along both its horizontal and its vertical axis, resulting in a concave shape interior surface and a convex exterior surface. The surface of copper alloy bowl was covered with copper and iron corrosion products. There was a missing section of rim on one side of the object, which was obscured by the presence of the concretion. One section of the rim was semi-detached and was split at the edge, filled with concretion and/or sediment. The majority of the concretion is thick and solid, adhering firmly to the bowl. The impression of the gun is quite clear (Fig. 3.27) but an area of concretion was soft and liable to loss.

Despite the degree of distortion, the form of the vessel indicates that the base was deep and rounded, suggesting a basin or bowl, rather than a plate. For example, this object exhibits a similar pattern of crumpling to the once rounded bridle boss shown in Egan, 2005, no. 1058. A form similar to the pewter surgeon's basin from the *Mary Rose* is a possibility (Jones, 2003, 85). It is less likely to

		Composition						
	Location	Cu%	Zn%	Pb%	Sn%	Ag%	Fe%	Other
Exterior Surface	Surface	69.2	26.0	0.3	2.6	0.2	1.6	Ni
	Base rivets	70.3	26.3	0.3	2.0	0.2	1.0	Ni
	Semi-Detached rim	66.0	27.5	0.4	2.9	0.2	2.8	Ni, Mn
	Rivet (uncleaned)	53.9	25.0	0.3	3.3	0.4	16.8	Ni, Mn, Zr
	Rivet (cleaned)	66.6	15.0	0.5	1.4	0.3	15.5	Ni, Mn, Zr
Interior Surface	Concretion	68.6	25.4	0.3	2.1	0.2	3.4	Ni
	Concretion	0.0	1.9	0.0	0.0	0.0	97.1	Co, Mn, Zr
	Semi-detached rim (left)	57.8	31.0	2.2	2.6	0.3	5.9	Ni, Mn
	Solid concretion	0.0	6.1	0.2	0.0	0.0	89.8	Co, Mn, Zr

Table 3.6 *pXRF composition of the copper alloy and concretion (GSP201.1). The high level of zinc and low lead levels suggest this can be categorized as brass*

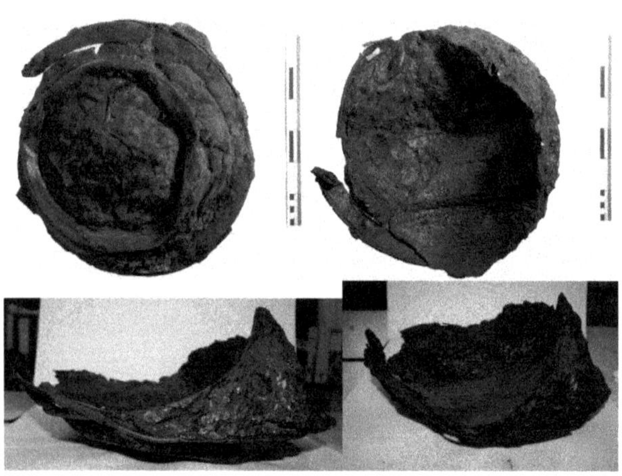

Figure 3.27 Copper alloy fragment and concretion GSP201.1 with impression of gun GSP201 visible on interior concave surface of concretion (photo Gresham Ship Project)

be a cooking vessel, such as a cauldron, as there is no evidence on this object of foot attachments or notches/attachments along the rim for a hanging mechanism (Gardiner and Allen, 2005, 432; Butler, *et al.*, 2009).

A riveted circle section of metal can be seen on the central base of the object (diameter of rivet outer circle: 57 mm). After chemical and mechanical cleaning the rivets on the rim and base were seen to be pink in colour, indicating a 'purer' copper composition, confirmed with pXRF analysis with a decrease in zinc content for these areas (Plate 6, Fig. 3.28 and Table 3.6).

The copper alloy metal is relatively thin (>5 mm) and likely to have been one piece of metal, set into a circular band of metal, the width of the present rim. The interior bowl edge has been curled over this circle of metal and the interior edge between the rim and the beginning of the wall of the vessel. The rim appears to be split into two sections, with the interior rim appearing to curl over the exterior sheet. Six rivets are evident on one section of rim and appear to continue along the semi-detached rim section, revealing the remains of the grommet in which the rivet would have been set. However, these do not penetrate into the internal face. These rivets do not appear elsewhere on the majority of the rim.

Conservation Treatment

One aim of the conservation treatment was to enhance the appearance of the object. This was achieved by revealing some of the brass metal surface below the corrosion product and by removing concretion around the rivets on the base and rim of the object. To present this as a 'shipwreck object', the retention of the concretion is significant and therefore the aim of the treatment was to retain as much of the concretion as possible. It was not intended to remove all of the corrosion products on the surface of the object, as this can provide support for the damaged structure of the vessel, which was particularly true for the more fragile rim sections, which were filled with concretion deposits. The calcareous deposit on the outer surface of the vessel could not easily be removed by mechanical cleaning. Therefore, dilute acid and chelating agent solutions were tested to determine an efficient cleaning method to remove corrosion from the copper alloy surface (Table 3.7).

Mechanical cleaning, aided by chemical poultices (30% w/v citric acid and 10% w/v EDTA disodium salt solution mixed into a 5% w/v Laponite RD in de-ionized water), revealed the bright golden yellow colour of brass metal just beneath the corrosion product (Huda, 2002). To avoid over cleaning the bowl surface, the application of the poultice was limited to 5 minute periods before removal and rinsing. The most efficient technique for removing calcareous deposits was found to be dilute citric acid applied as a poultice (Laponite RD gel) to loosen the deposits, followed by mechanical removal and then rinsing with de-ionized water.

A two-phase desalination treatment was carried out using tap water followed by de-ionized water immersion. This was monitored weekly with standardized conductivity,

Citric Acid 30% w/v in deionised water (to remove the calcareous deposits)
Ethylenediaminetetra acetic acid (EDTA) disodium salt 10% w/v in deionised water (a chelating agent for ferrous metal ions)
EDTA tetrasodium salt 30% w/v in deionised water (a chelating agent for calcium)
Sodium hexametaphosphate (i.e., Calgon™) 30% w/v in deionised water (a chelating agent for calcium carbonates)
Laponite RD (synthetic inorganic clay, [magnesium silicate]) 5% w/v in deionised water (a gel poultice for the application of citric acid 30% v/v in deionised water)

Table 3.7 Chemical Cleaning Tests for copper alloy/ concretion (GSP201.1)

was monitored weekly with standardized conductivity, pH, Cl, Fe, silver nitrate (chloride presence) measurements. Tap water was changed weekly until conductivity measurements remained unchanged for one month. This process took six months, prior to starting desalination with de-ionized water (Plate 7, Fig. 3.29). Following the completion of desalination and drying, the object will packed within a desiccated enclosure, designed to stay below 35% Rh.

Summary of Small metal finds

The decorated objects, tin alloy salt holders (GSP86.1, GSP14) and spoon (GSP86.7), silver spoon (GSP20.1), copper alloy jug (GSP208) and copper alloy bowl (GSP201.1), provide a small but interesting group of finds. These have been documented, described, cleaned and stabilized to present conserved objects in various states. The chemical cleaning of the tin alloy objects and silver spoon aimed to reveal surface detail from beneath the marine deposits formed on the objects. The outer metal surface of the copper alloy bowl was revealed, whilst attempts were made to retain the large iron concretion adhering to the inside. Different technical solutions were applied that combined robust mechanical removal techniques with more subtle uses of chemical cleaning that were adapted to the particular requirement of each treatment. The excavation and conservation process has provided raw material for further research. The small number of objects in this group does restrict our ability to develop interpretations concerning shipboard society, but they do have the potential to tell interesting stories and become powerful museum exhibits.

The association between the ceramic fragment rim that adheres to the silver spoon (GSP20.1) provides an intriguing object that hints at the use of the spoon in board the ship. We could speculate that the spoon was scooping contents of the ceramic bowl to which it is now firmly attached. We can however be more definitive in that we can say that both the spoon and the ceramic, the copper alloy bowl GSP201.1 and the gun GSP201 were joined together as shipwreck objects and have now been stabilized in this association as part of the long term archaeological archive. The long term associated between the spoon, the ceramic, the copper alloy bowl and the gun present intimate relationships that reflect the wrecking event and the taphonomic processes that led to the survival of these objects in this form. These objects have been transformed by their association with the wrecked ship and their time as part of the marine environment, prior to the processes that lead to their recovery and most recently by their transformation as conserved archaeological objects.

Organic Small Finds (personal possessions)

Small Wooden Finds

Six small wooden finds had been retained as part of the finds assemblage. These included a pike shaft GSP204 (Fig. 3.30 and Plate 7, Fig. 3.33), barrel stave GSP11 (Fig. 3.32), trenail WA 54135 and three small objects GSP32 (Fig. 3.31), GSP82 and GSP207 (Hildred, 1997). Each object was assessed to determine the research potential of the material for the Gresham Ship Project. The condition of each object was documented and species of wood identified prior to designing an appropriate conservation treatment to stabilize the wood finds for archive deposition.

Condition assessment

Detailed documentation (photography, illustration, dimensions, X-radiography and species identification) was conducted in order to assess the treatment regime needed for each wooden artefact. Drawings noted the transverse, tangential and radial planes in order to assess impregnation rates and potential shrinkage during treatment. All six fragments were in a solid dense condition, although the pike GSP204 exhibited severe marine mollusc attack at one end (Fig. 3.30). Dense calcium linings visible in X-radiography indicated borer tunnels had penetrated one-third of the pike's length. The pike tip also exhibited iron staining, resulting from the associated iron tip (part of GSP205: see Plate 11, Fig. 3.44). A small wedge fragment GSP32 exhibited heavy iron corrosion staining and associated concretion from an iron pin which, when viewed as thin-sections under transmitted-light microscopy, appeared to have partially mineralized the wood (Fig. 3.31) (Blanchette and Hoffmann, 1994).

Species identification

Species identification and cell wall condition was assessed, using thin sections under light transmitted microscopy. This was required in order to assess impregnation rates and molecular weights of polyethylene glycol needed for treatment (Cook and Grattan, 1991). Five samples were identified as oak (*Quercus* spp.) which is known for problems associated with consolidation using polymers with large molecules, cross checking on radial surfaces and wide cracks in tangential surfaces post treatment (pers. comm., Jacqui Watson).

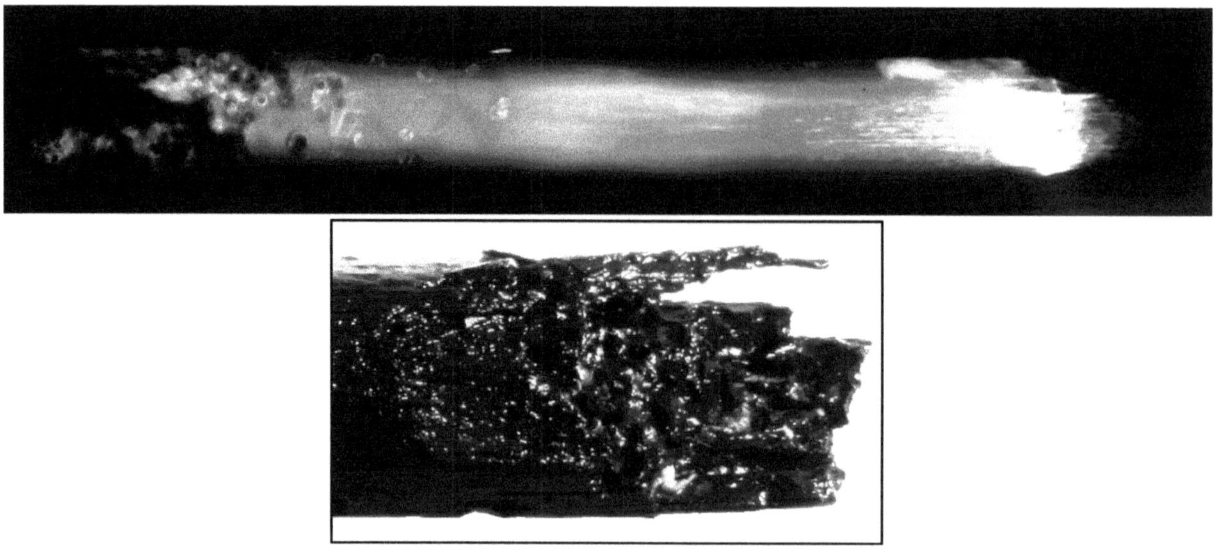

Figure 3.30 *Pike shaft GSP204 before conservation: above woodborer activity visible in X-radiograph; below holes created by marine woodborers; white areas are calcareous material they left behind (photo Gresham Ship Project)*

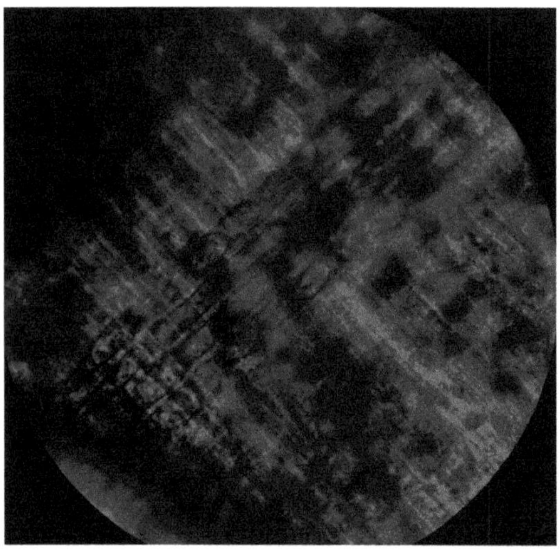

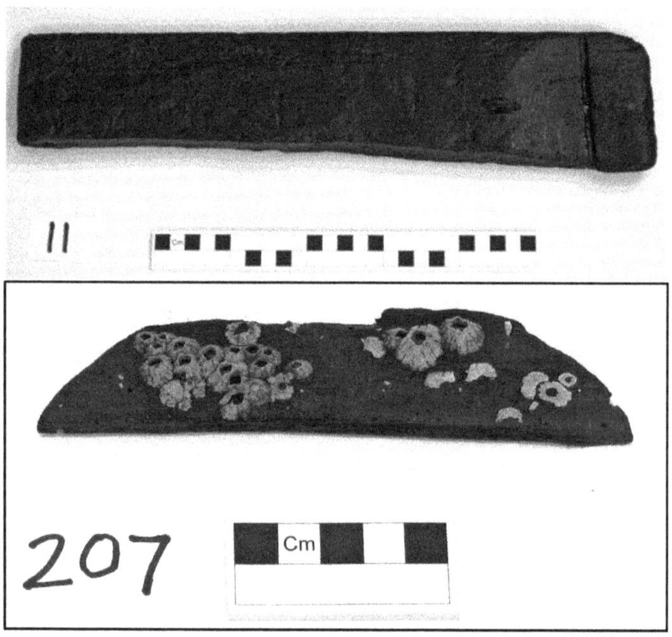

Figure 3.31 *Left GSP32 photomicrograph of ring porous cell arrangement showing intact early wood cells (TS x 40). Cell lumen filled with iron corrosion products (photo Gresham Ship Project)*

Figure 3.32 *Right barrel fragments before conservation: above oak barrel stave GSP11; below pine barrel lid GSP207 with marine growth (photos Gresham Ship Project)*

The small fragment GSP207, possibly part of a barrel lid, was identified as pine (*Pinus* spp.), a species which responds well to Polyethylene Glycol (PEG) impregnation and freeze-drying due to its simple cell structure (pers. comm., Jacqui Watson).

All fragments underwent desalination in tap water over a period of four months to remove the bulk of soluble marine salts. Objects were stored at 4°C to prevent bacterial degradation. Conductivity levels of bath water were monitored using an InoLab pH/Cond 750 meter. Tap water was changed when levels stabilized and desalination was completed when conductivity levels reached the same as tap water.

Treatment

Iron staining posed several problems for the pre-treatment and post-treatment stability of the wood. Potential problems include physically blocking the microstructure of wood, making it impermeable to bulking agents; physical degradation caused by oxidation of iron sulphides into sulphuric acid and structural damage caused by crystallization of iron pyrite crystals (Watson, 1985, 213, 217; Jones, 2003, 63). Ferric (Fe^{2+}) ions from the burial environment are actively chelated (bound) by cellulose and tannates that build up over time. Under anoxic conditions, iron salts are converted into sulphides (pyrite) by sulphate

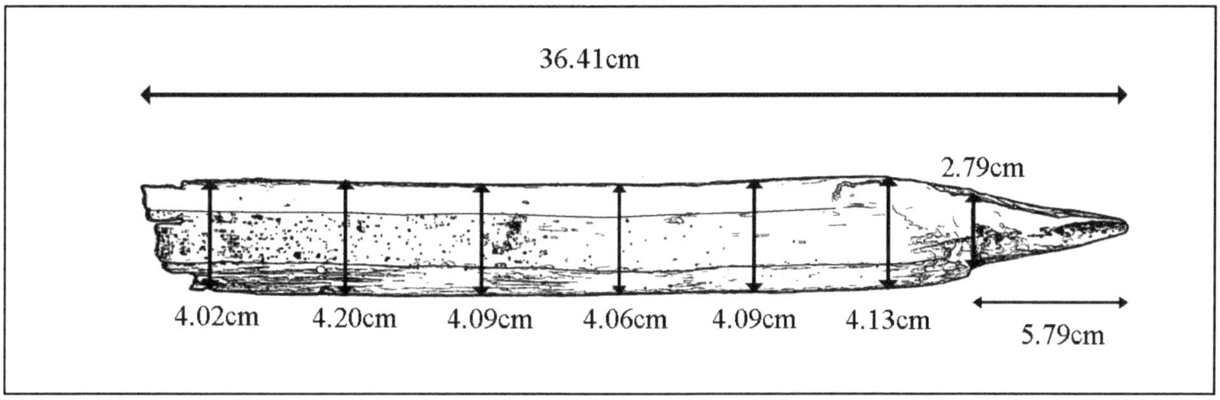

Figure 3.34 Pike shaft GSP204: pre-treatment measurements

reducing bacteria. Iron sulphide is unstable in the aerobic environment and its oxidation can destroy an object. Sulphuric acid produced inside the object will start to hydrolyse the wood (Watson, 1985, 213; Jones, 2003, 63). Pyrite is also reduced to gypsum by oxidation or bacterial action; the crystallization of gypsum can cause distortion of the cell wall (Watson, 1985, 213, 217). High concentrations of iron salts (iron pyrites) in dried archaeological wood results in structural damage and blooms of sulphate crystals occur on the wood surface (Jones 2003, 63). Removal or stabilization of iron salts needs to be achieved to mitigate damage (MacLeod, *et al.*, 1994; MacLeod and Richards, 1997). The use of a complexion agents such as EDTA (disodium salt of ethylene-diaminetetraacetic) and DTPA (diethylenetriamine-pentaacetic acid) were considered as a method to reduce residual insoluble iron salts (Fe^{3+}) to soluble salts (Fe^{2+}) prior to bulking. Removing iron salts is, however, difficult, as many iron compounds are insoluble in most chelating agents and can be only dissolved by strong mineral acids, which have been found to soften wood cell walls in wood and lead to increased shrinkage in waterlogged leather (Watson, 1985, 213; Jones, 2003, 63, Karsten and Graham, 2011). Therefore, no attempt was made to remove ferrous ions from the waterlogged organic materials, except where this was considered necessary to reveal surface detail from beneath obscuring deposits (for example, leather shoe GSP35) (Hovmand and Jones, 2001; Hovmand, 2002).

Due to the small size of the wooden objects, it was not possible in most cases to carry out destructive condition assessment measurements, such as maximum moisture content (µmax) (Cook and Grattan, 1991, 245; Panter and Spriggs, 1997, 188). Concentrations and impregnation time were therefore determined by other means, such as visual assessment, X-radiography and optical microscopy, the pin test and comparison with previous conservation research. These suggested variations in the treatment regime from 10–25 v/v % Peg 400 in water and 0–20% w/v Peg 4000.

Pike shaft GSP204

GSP204 (dimensions 364 x 40 x 42 mm) was identified as the tip of a pike shaft (Fig. 3.33 on Plate 7 and Figs 3.30 and 3.34). Thin-section microscopy indicated that the wood is ring porous with wide multisereate rays, identified as oak (*Quercus* spp.; pers. comm., Jacqui Watson and Martin Bridge; Hather, 2000, 46). The pike has been manufactured from the length of the tree, its cross-section corresponding to the transverse section, providing the object with maximum strength. The pike is octagonal in cross-section and the spike-end has been worked in five-planes. There are traces of residual iron at the spike end and a fibrous material is adhered to the spike end and is possibly the remains of caulking or marine growth.

Similar pikes were used commonly in medieval European warfare from the 13th to the 18th centuries after which the widespread use of muskets rendered it obsolete. A Tudor pike would have ranged from between 3 and 6 m in length. On board ship, they could be used for boarding or to repel boarders. The long length of this weapon required a very strong wood, such as well-seasoned oak or ash and the shaft would be tapered towards the end to prevent sagging (Brown, 1997; Hather, 2000).

Pre-Treatment Condition

The pike end is complete in shape, the outer surface of which was soft and the core dense. The pike end was much denser than the shaft end, possibly due to mineralization of the wood by iron corrosion and absence of borer activity. White marine coral concretions were present in small areas at the shaft end, associated with large marine borer tunnels (Fig. 3.30). Several small fragments were detached from the shaft end, which was fragile and liable to loss during handling.

Optical microscopy of thin sections taken from the surface show cell walls to be intact and lumen to be filled with dark red iron corrosion products (pers. comm., Jacqui Watson). The availability of small samples of detached wood enabled maximum water content (µmax) to be tested. This suggested a µmax of 192%, indicating that the wood was in reasonable condition (Grattan and Clarke, 1987). Drying treatments for waterlogged wood can easily result in significant dimensional changes; therefore, the dimensions of this object have been accurately recorded before treatment (Figure 3.34).

The object was desalinated by placing in fresh tap water and measuring salt levels with a conductivity meter until

Figure 3.36 Rope fragment GSP211: left before treatment; right after treatment (photos Gresham Ship Project)

they matched that of the tap water. A treatment regime 20% v/v PEG 400 and 20% w/v PEG 3350/4000 was calculated using the PEGCON programme (Grattan and Clarke, 1987, 178). The concentration of the PEG 4000 solution was increased in 5% increments over several months. Impregnation was monitored using a hand held refractometer and an associated calibration curve was used to measure the concentration of PEG solutions and manage increments in PEG concentration. After a month in the final PEG solution, the object was taken out, excess PEG was removed from the surface and it was frozen ready for freeze-drying.

Hammer GSP202.2

A similar treatment regime was adopted for wood/iron hammer GSP202.2 (dimensions 107 x 120 x 35 mm). Desalination took place in tap water with 1% v/v triethanolamine as a corrosion inhibitor to help protect the ferrous metal elements. The object was impregnated with 20% v/v PEG 400 and 20% w/v PEG 3350/4000, with the addition of 1% v/v triethanolamine, (Plate 8, Fig. 3.35). The iron hammer head was actively corroding and had lost significant material in the cheeks areas. The wooden handle was encrusted and impregnated with iron corrosion products. The juxtaposition of iron/wood as composite objects creates additional problems in the conservation of waterlogged marine objects. PEG-treated wood associated with iron component parts in a composite objects can result in reciprocal damage in the long term (Jones, 2003). This is only partially mitigated by the addition of corrosion inhibitors during the impregnation of PEG. This, however, was not considered sufficient justification to separate the ferrous metal hammerhead from the remains of the wooden handle for separate treatment (Krop and Nordgren, 2011).

The hammer, although a part of the concretion (GSP202) that contained the powder chamber (GSP202.1), could not have been effectively stabilized as part of the concretion. Therefore, a conscious choice was taken to physically separate the wooden handled hammer from its concretion, so that conservation could be carried out. This does, however, diminish the connection between the hammer and the powder chamber. The intellectual and physical association can, of course, be regained after conservation by reuniting the separated elements. Therefore, the archaeological provenance information that recreates the association between the objects becomes a critical tool to reunite dislocated materials.

Physical separation as a result of the archaeological recovery process does routinely occur and was an advantage for the conservation treatment of the pike shaft (GSP204). This allowed the PEG impregnation and freeze-drying treatment to be designed specifically in relation to the condition of the wooden shaft. The treatment of the associated shoe sole (GSP205) attached to wood and the iron pike head was, however, a compromise in order to balance the conservation needs the wood, leather and iron components. The choice of a two stage PEG impregnation treatment followed by freeze-drying was selected for GSP205 in order to conserve the wooden elements of the object. This treatment has some benefit for the leather element (although the use of PEG 3500/4000 is not normally considered appropriate), but provides little benefit of the iron pike tip. This is because the use of PEG and the freeze-drying can have an adverse effect on the chemical and physical stability of the iron. In other words, the treatment of composite objects requires compromises, as the work on shoe concretion GSP88 also demonstrates.

Other Organic Materials

Sections of rope fragments were associated with several concretions, as likely to be part of the rigging entangled with heavier objects as the ship sank, or during the time on the seabed prior to concretion formation, or associated with the salvage operations in 1846, as they are to be associated with the particular object during the life of the ship. A fragment of rope GSP211 was connected to the concretions around gun GSP201 and was, therefore, also associated with copper alloy bowl GSP201.1 and ceramic jar GSP201.2. The rope was separated as a discrete find immediately following excavation and was therefore treated as a separate object (Fig. 3.36). It was identified as a z-spun, S-ply, Z-cabled (zS2(Z)3) rope (Wendrich, 1999, 31). Fibre identification was attempted using polarized light microscopy (PLM) in order to determine whether the rope was hemp or flax or later materials such as jute (that might occur from 17th century onwards). The degree of deterioration meant that

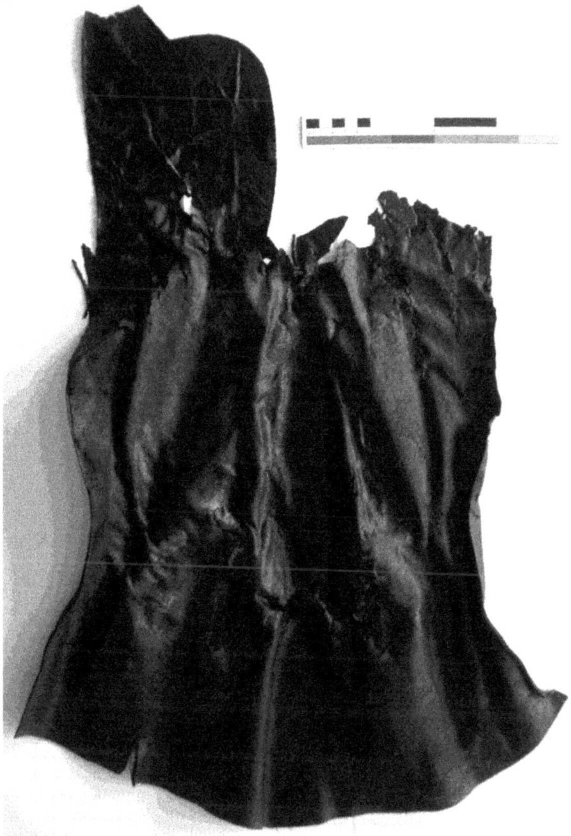

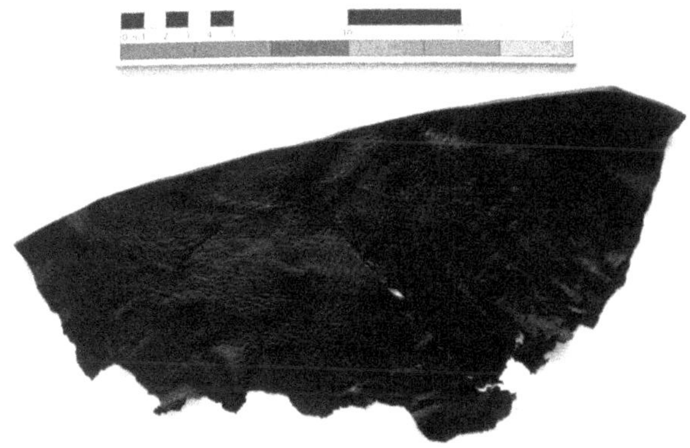

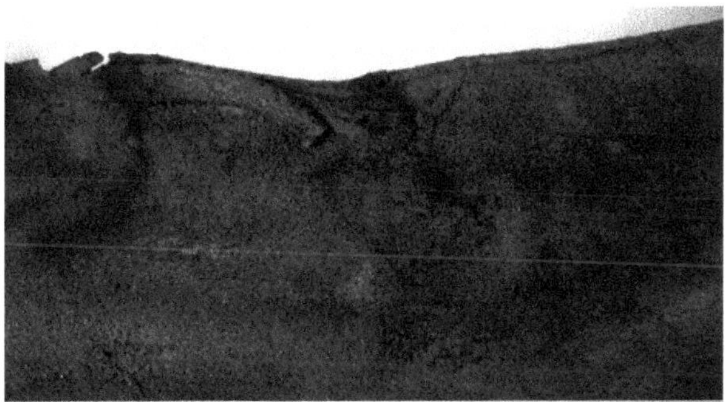

Figure 3.38 Leather boot fragments: above large piece of leather GSP80 part of a boot upper; top right smaller fragment, possibly from second boot (photos Gresham Ship Project)

Figure 3.39 Leather boot GSP80: centre right detail of tunnel stitch on inside surface, where bootstraps may have been attached; bottom right detail of stitch holes from back leg seam, with larger holes from heel attachment (photos Gresham Ship Project)

diagnostic features (plant cells and crystalline features), could not be clearly identified (Catling and Grayson, 1982; Ballard and Skals, 1996). An attempt was made after further cleaning (ultrasonic cleaning) and solvent drying. However, the diagnostic features were obscured by deposition of the thick organic matter, so no further identification was possible (Garside and Wyeth, 2006). The rope fragments were solvent dried in a sequence of solvent baths; 50% v/v de-ionized water/Industrial Methylated Spirits, (IMS) (95% ethanol/5% wood naphtha), 100% IMS, 50% v/v solution IMS/acetone and 100% acetone. This process was monitored using hydrometers to determine when the object was in equilibrium with the solvent bath, after which the object was air-dried. The dried rope was consolidated with Klucel G (hydroxyl propyl cellulose) 0.5% w/v in IMS (Florian, *et al.*, 1990; Wevers, 1991; Peacock, 1990; Smith, 2003; Huisman, 2009). After treatment the rope was slightly stiff, friable and had shrunk in lenghth by 20% . It was however, easier to handle and is suitable for further study and future display.

Leather footwear

A small group of leather objects formed part of the finds assemblage; the remains of two leather shoes (GSP55 and GSP88), three shoe soles (GSP35, 205 and 206) and part of what was initially described as a leather garment or jerkin (GSP80: see Fig. 3.37 on Plate 9 and Fig. 3.38). The identification of GSP80 as a boot upper, allows us to discuss the footwear collection from the ship, representing six items in total. Investigative conservation procedures were performed in order to discover information about the footwear and potentially the wearers (van Driel-Murray, 1987; von Brandenberg, 2009). Each treatment was carried out by a different graduate conservation student and therefore the approach to each object reflects the individual choices made by each conservator. It is therefore appropriate to describe in detail the treatment process of each object and then to reflect how this relates to the overall aim of the conservation process to conserve the objects to reveal archaeological information about the wreck assemblage, whilst seeking to present the finds as shipwreck objects.

Over-the-knee boot GSP80

Four pieces of vegetable tanned calf leather comprising one piece (dimensions 630 x 500 x 5 mm) and three smaller fragments were initially thought to represent the remains of a jerkin (Frith, 2006), but were subsequently identified by their size, shape and the positioning of stitch holes to be the upper of an over-the-knee boot (Fig. 3.38 and Plate 10, Fig. 3.40). One of the smaller fragments may have come from the matching boot – it was full thickness and had a cut edge. The other two fragments were delaminations, possibly from the larger piece.

The boot upper appears to have been cut from a large piece of leather to a standard pattern, but with the owner's size and shape in mind (Goubitz, *et al.*, 2001, 31). The line of the back seam follows the profile of the back of the leg, indicating a tight fitting style, with straps on the inside and a peak at the centre front. The main seam down the back of the leg would have been closed with the edges turned inwards, secured with stitches placed 2 mm apart. Although no remains of the stitching was present, it is likely that this would have been flax or hemp thread (Kite and Thomson, 2006, 246). Marks on the flesh side, near the seam, are the remains of tunnel stitching, presumably used to attach a strap with which the wearer would pull the boot on (Fig. 3.39). There are five large holes cut out of the leather at the heel end, these large stitches would have secured the layered leather to create the heel. There is no evidence that the boot was lined.

The additional workmanship and materials meant that boots of this kind were likely to have been reserved for higher-status crew members, if only due to their cost. There are fewer examples of archaeologically recovered leather thigh boots than of other types of footwear (Gardiner and Allen, 2005, 84). A reason for the unrepresentative ratio of boots to shoes found is likely to be that uppers were reclaimed for secondary use after the shoe/sole elements were worn out. It is for this reason that shipwrecks tend to yield more identifiable fragments and complete shoes/boots than land sites (Goubitz, *et al.*, 2001, 230).

Treatment

From the time of its recovery until its treatment in 2009, GSP80 was stored rolled in a polythene bag filled with tap water (Plate 9, Fig. 3.37a, top left). The leather was heavily stained with iron corrosion products, covered with silt and mud and was distorted, where it had been folded and rolled within the storage bag. Over half of the large piece, towards the lower end of the leg where only one ankle flap remained, was subject to delamination. The leather was cleaned under gentle running water with a soft brush to remove silt, mud and iron corrosion products (Jennsen, 1987, 129). During cleaning, a polythene tracing was made of the shape of the large piece, marking the position of stitch holes. From this a wool felt replica was constructed to examine the three dimensional form of the object.

The object was desalinated by soaking in fresh water until conductivity of the water was stable, followed by impregnation with of 20% v/v glycerol (1, 2, 3-propanetriol) solution in tap water for three days (Ganiaris, *et al*, 1982; Wouters, 1986; Suenson-Taylor and Sully, 1997; Peacock, 2001; Kite and Thomson, 2006, 248). The three fragments were mounted around an inner support of a stiffened textile and acid-free tissue to maintain the rough shape of a boot, equivalent to the replica. The outside was wrapped with nylon gossamer and secure with cotton tapes. Prior to freezing the object was patted dry with paper towel and freeze-dried at 20°C for two weeks (GSP80, GSP55 and GSP206 were all impregnated with glycerol and freeze-dried in a similar way). The leather dried with minimal shrinkage and remained flexible. The stitch holes appeared to be more clearly defined after drying (Fig. 3.39 and Plate 10, Fig. 3.40). The boot was stored with padding to maintain the curved shape and take the weight of the leather away from the fold.

Shoe concretion GSP88

Shoe concretion GSP88 (dimensions 460 x 180 x 130 mm) was a composite object consisting of concretion, wood and leather and was the most complicated of the leather finds from the Gresham Ship (Plate 10, Fig. 3.41). The leather was partially covered by hard concretion. The interior of the shoe was filled with pieces of concretion and the concretion at the heel end of the shoe was attached to a wooden fragment 430 mm long. The whole object is in relatively poor condition and vulnerable to further damage due to the association of these different materials. The concretion is heavy and liable to oxidation, the waterlogged wood is swollen and liable to shrinkage on drying; likewise the leather is brittle from its mineral content and liable to physical damaged due to the 4 kg weight of the concretion.

Much of the shoe was enclosed in hard concretions, delaminating in areas with the heel and toe both obscured. The remains of this adult-sized leather shoe consisted of the sole and parts of the upper of a right shoe. It appeared to be a simple construction with a single piece vamp, which may have been a slip on or had latched fasteners on the sides (pers. comm., Jackie Keily). Most of the sole of the shoe was visible and three layers could be seen in profile, which are probably the insole, midsole and treadsole (made in two parts). There was a repair on the toe area of the sole and the sole had been distorted inwards in the middle, which may have occurred during the wrecking event. The post-1500 AD date for this type of shoe construction fits with the late 16th-century wreck, but does not provide additional dating information for the ship's assemblages (Goubitz, *et al.*, 2001).

Treatment

The object was cleaned by gentle brushing under running tap water. Some of the corrosion products were loosened and removed during cleaning, but the majority were too hard to be easily removed. The object was desalinated by immersion in tap water baths, which were changed daily until a consistent electrical conductivity reading close to that of the tap water was reached. The object was impregnated up to a final concentration of 20% v/v PEG 400, 20% w/v PEG 4000, with 1% v/v triethanolamine (corrosion inhibitor) solution in tap water, prior to freeze drying at -20°C for several weeks.

Shoe fragment GSP35

Shoe fragment GSP35 was the leather sole of a right shoe (dimensions 195 x 65 x 4 mm). Little of the leather surface was visible, as the sole was heavily impregnated and encrusted with iron corrosion products, giving it an ochre colour and obscuring any surface details of the leather (Plate 11, Fig. 3.42). The oxidation of iron salts within the leather is likely cause both physical and chemical damage to the object. The object is particularly degraded near the centre of the sole. The weakening of the leather in this area suggests evidence of use, as it is where the metatarsal head or ball of the wearer's foot would be when wearing the shoe. X-radiography confirmed the absence of any metal fittings on the object and showed an unremarkable mottled appearance in areas of the sole covered by iron deposits.

Treatment

Prior to cleaning, the object was desalinated and then soaked in 5% w/v disodium EDTA (ethylene-diaminetetra-acetic acid disodium salt) in water for 23 hours. This was designed to soften the obscuring layers of and encrusted with iron corrosion products. The bulk of softened accretions were mechanically removed with toothbrush and bamboo stick, then rinsed thoroughly with tap water for two days. The object was impregnated with glycerol and freeze-dried. Some warping of the sole had occurred during freeze-drying, which was corrected by re-humidification. A backing support was added to the thin, vulnerable areas of the shoe sole, using tabs of Japanese tissue with Mowilith 50 (polyvinyl acetate) 50% w/v in acetone across the join; then a larger patch of watercolour-tinted Japanese tissue was applied over the area of the repair, with 1% w/v Klucel G (hydroxy propyl cellulose) in IMS. The repairs were not visible from the front of the sole and provided enough support to reduce the risk of breakage during handling. Stitch holes were uncovered with removal of the accretions and can be seen around the perimeter of the sole, but there was no evidence of stitching thread was present (Plate 11, Fig. 3.42).

Heel fragment GSP55

GSP55 consists of fragments of a shoe upper attached to a concretion (dimensions 138 x 210 x 60 mm). The remains of the leather shoe consist of a fragmentary sole and heel (heel stiffener), with insole, outer sole, welt and right quarter evident. There is a free standing seam with two rows of stitch holes, characteristic of welts (Plate 11, Fig. 3.43). The insole and outer sole have evidence of stitches. A heel stiffener was typically a triangular piece of leather that was sewn into the heel section of the footwear to strengthen an area that was subject to great stress and wear (Grew and Neergaard, 1988). The quarter is one part of a two part type as it has evidence of being gathered through stitching in the middle of the heel (Evans and Mould, 2005). The bovine leather is delaminated, fragmentary, torn and very fragile, within some impregnation of iron corrosion products.

The shoe was connected to the remains of a concretion that was initially connected to one of the iron guns recovered from the wreck (Plate 9, Fig. 3.37, bottom left). The concretion is broken into two fragments and appears to be holding parts of the leather shoe together. The shape of the concretion and associated impressions of leather suggest that the tip of the shoe was attached to the concretion at some point.

Treatment

The object was cleaned in running water, desalinated in repeated baths of tap water for four days and monitored until conductivity reduced to that of tap water. The object was impregnated with 20% v/v glycerol (propane-1, 2, 3-triol) 1% v/v triethanolamine and freeze-dried at -20 °C for 2 days. A backing support was added to repair vulnerable areas using tabs of Japanese tissue with Mowilith 50 (polyvinyl acetate) 30% w/v in acetone across the join; then a larger patch of watercolour-tinted Japanese tissue was applied with 1% w/v Klucel G (hydroxy propyl cellulose) in IMS.

Leather shoe GSP205.

This leather outer sole GSP205 (dimensions 280 x 140 x 60 mm) is attached to a wooden timber from the shipwreck and iron pike head, which connect to pike shaft GSP204. The sole is nearly complete; the only areas of loss at the toe of the sole and along the inner edge could be due to original wear. The shoe is late 16th-century in date, vegetable tanned bovine and equivalent to a modern UK adult size 7 (Grew and Neergaard, 2001, 102). The shoe is of welted construction, seen in the punched stitch holes around the perimeter of the sole (Plate 11, Fig. 3.44). There are multiple punched holes associated with the areas of wear, which may indicate a past repair (pers. comm., Jackie Keily). The corrosion from the iron pike head has affected the adjacent areas of the shoe sole.

Following desalination the shoe was impregnated with PEG 20% v/v polyethylene glycol 400, 1% v/v triethanolamine (corrosion inhibitor) solution in tap water. Once freeze-dried, iron pike head will be coated in Paraloid B-44 (a methyl methacrylate copolymer) with a top coat of microcrystalline wax.

Leather Shoe 206

Shoe 206 (dimensions 116 x 91 x 265 mm) is possibly the forepart of a left shoe's leather sole or a midsole or insole, but more likely a repair (Goubitz, *et al.*, 2001). The leather has been cut to shape and the holes left from stitching show that all of its sides would have been stitched to the shoe. There is evidence of use in that this sole is thinner near the top of the shoe where the wearer's toes would have worn down the leather (Plate 12, Fig. 3.45). Although the toe of the sole is worn, it is likely, from the rest of the sole, that it would have originally been rounded in shape, late 16th century in style, most likely to be from a slip-on, everyday work shoe (pers. comm., Jackie Keily). It is possible that the sole was added as a 'clump sole', a term describing a half sole added to a shoe as a repair or simply thicken the sole (Carlson, 2005). In the case of shoe 206, it was stitched on the inside to strengthen the sole at an area of weakness. This is suggested as the back of the sole is cut diagonally, so it would not have covered the front area of the sole completely, as you would expect if it were added to thicken the whole sole. The follicle pattern is not discernible under high magnification due to abrasion; however, the leather is likely to be bovine leather due to its thickness and the context of the object. As this sole was excavated from beneath GSP205, it is likely that the two were associated, with GSP206 being a repair to GSP205. The archaeological process that allocated separate object numbers to these objects has been emphasised by the separate conservation treatments that each object has undergone.

Treatment

The object was desalinated and then immersed in a 10% v/v glycerol/10% v/v PEG 400 solution, freeze-dried for two days until a constant weight was reached. The treatment was successful, as the object is now stable enough to be handled, examined, transported and displayed. The object can now consequently fulfil its research, educational and outreach potential. The object did suffer inevitable dimensional change as a result of drying, but this was deemed worthwhile, as drying has made the object stable and understandable.

Summary of Organic Small Finds

It is clear from the differing responses to the conservation of the six leather footwear objects that the outcomes of the conservation process are not readily predictable. Despite a group of similar objects conserved within an overarching aim, the outcomes of each treatment differ. The decision-making process associated with these outcomes can be considered in relation to the two extreme examples of how the conservator may alter the physical fabric of the artefact, to achieve a conserved object in its manufactured form or as a shipwreck artefact as found.

For objects GSP80 and GSP206, a stable, legible, conserved object was produced by applying a standard interventive conservation process of desalination, impregnation, freeze-drying and post-drying repair. This is a consequence of the condition of the object and the availability of effective conservation treatments to investigate and stabilize objects of this type. GSP88, GSP55 and GSP205, however, involved technically more complex conservation treatments, due to the presence of composite materials (leather, wood, ferrous metal, concretion), but also due to the desire to retain as much of this material as possible in the conserved object. This limits the shared goals of investigation and stabilization, which in the case of GSP88 in particular, have been compromised by the desire to retain the concretion as part of the conserved object. For GSP88, the aim to produce a conserved object that would be able to communicate effectively as a shipwreck artefact, has resulted in limited archaeological information about the shoe itself, as well as in a conserved object that is vulnerable to physical and chemical change over time. A different approach was taken for GSP35, where the conserved object was altered from the object as found. The desire has been to remove the contained iron salts in order to ensure the object's long-term stability. The chemical removal of surface accretions was justified in order to satisfy the requirements of investigation and stabilization.

Monitoring the condition of the hull timbers

The most significant waterlogged organic feature is, of course, the ship itself. The relocation of the hull timbers to Horsea Lake and finally to Stoney Cove, provided an opportunity to research the extent and rate of deterioration of the oak-built hull subject to a fresh water oxygenated environment. A repeatable methodology for the condition assessment of timber samples from the ship was developed to monitor the changes that occur over time as a result of this change of environment. This has involved an examination of agents of deterioration and their impact on oak cell structure, in order to understand the mechanisms of decay affecting the ship's hull. The condition of the hull was quantified through a range of chemical and physical analytical techniques, applied to sample material removed from the ship. From this, a standard methodology was developed to assess the condition of samples of the hull over time. This has provided a means of monitoring ongoing changes to the condition of the timbers that will inform any conservation interventions to stabilize the ship timbers. This will be the focus of a forthcoming conservation journal article (Sully *et al.*, forthcoming).

Ceramics and Glass

Two ceramic fragments (GSP31 and GSP34), one glass fragment (GSP58) and one ceramic/concretion (GSP201.2), underwent active conservation (Fig. 3.46). This provided an opportunity to assess appropriate desalination and drying methods for low-fired, high-fired, glazed and non-glazed ceramics and glass finds recovered from a marine environment. The aim of the conservation was to desalinate and stabilize for inclusion in the archaeological archive.

Condition assessment

GSP31 is a fragment of a red, low-fired ceramic. Visual examination of the cross-section indicated that the fragment had been fired in a reducing atmosphere followed by an oxidizing atmosphere, as shown by the colour variation in cross-section (grey central section and red outer edge). The fragment had been stored in tap water in a polythene bag prior to treatment. The surface was soft and easily scratched with a scalpel, particularly the cross-section.

GSP34 is a fragment of a high-fired white ceramic with grey inclusions and a red coloured glaze. Examination under optical microscopy indicated that the body was in a stable condition with glaze intact. The entire inner surface and cross sections were covered with strongly adhered marine incrustation ranging between 1 mm and 4 mm in depth. Exoskeletons within the encrustation

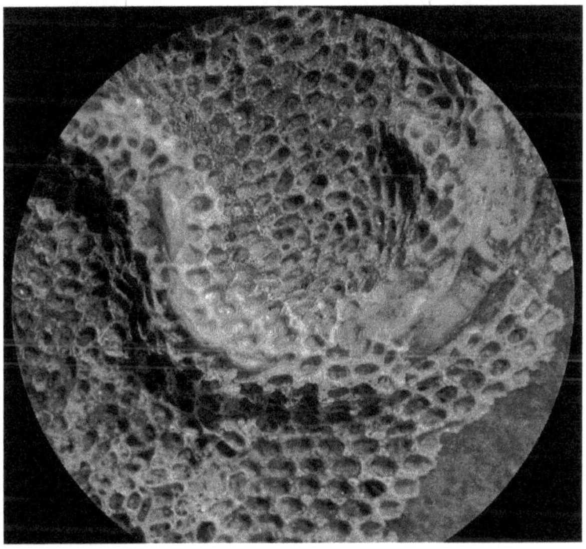

Figure 3.46 Ceramic and glass objects during treatment top left GSP31; top right GSP58; bottom left GSP34; bottom right detail of exoskeleton within marine encrustation adhered to GSP34 (photos Gresham Ship Project)

were visible under optical microscopy (Figure 3.46 bottom left). The glaze exhibited star-shaped fractures throughout, while red spots of iron corrosion occurred throughout the glaze.

GSP58 consisted of a base fragment of a free blown cylindrical straight-sided brown/blue glass bottle with pushed in dome-shaped indented base. Examination under optical microscopy indicated the glass to be solid and not liable to fracture. The outer surface exhibited scratches, chips and thin iridescence throughout. Break edges and the inner surface were weathered to a lesser extent than the outer surface and there were no residues visible within the bottle interior. Since it dates to the late 18th century or early 19th century (pers. comm., John Shepherd), it may have been thrown overboard from a passing ship or fishing boat, but there is the distinct possibility that it came from the surface support vessels of the 1846 salvage operation. The stoneware fragment (GSP34) and rope fragments (for example GSP211), might also be associated with this event.

Treatment

Desalination followed by controlled air-drying was considered to be appropriate treatment options for ceramic GSP31, GSP34 and glass object GSP58. These objects were desalinated in separate containers in the following stages: immersion in 100% tap water, immersion in 50% tap water and 50% de-ionized water, immersion 100% de-ionized water. The conductivity of each bath was monitored using an InoLab pH/Cond 750 meter. Bath water was changed once conductivity levels had stabilized. Objects were moved to the next stage once conductivity reached the same levels as the original bath solutions. Each object was air-dried in ambient atmospheric conditions and regularly monitored for salt crystallization. After 30 days objects showed no signs of salt crystallization or fracture indicating that desalination had been completed successfully. The first stage of

Figure 3.47 Fragment of Spanish olive jar GSP215: left showing weathered pine resin on inside surface; right detail of resin deposit (photos Gresham Ship Project)

desalination in 100% tap water was the most time-consuming as each object took between one and two months to reach the same conductivity levels as tap water. The subsequent stages were completed in one month. The process might have been speeded up, if the bath water had been changed more regularly.

Analysis

A layer of weathered resin was identified on the inside surface of ceramic fragment GSP215 during investigative cleaning (Fig. 3.47). The fragment had previously been dried and identified as a piece of a Spanish olive jar by Wessex Archaeology.

Resin is known to be used as a sealing agent to prevent jar contents from seeping into the porous ceramic fabric (Smith, *et al.*, 1998, 129). Pine resins (*Pistacia* spp.) have been recorded as being commonly utilised for this purpose and jars with resin probably carried liquid such as vinegar, wine or olive oil (Mills and White, 1989, 37; Beck and Borromeo, 1990, 51; Marken, 1994, 106).

The resin layer was sampled and analysed using FTIR and the resulting bond peaks were compared with several species of pine-resin from the Infrared and Raman Users Group online spectral library www.IRUG.org). Comparative peaks were observed in all pine resin samples of different sub-species in the 1457, 1384, 1246 and 1181 wavenumber (cm-1) regions, indicating the sample consisted of pine resin (Fig. 3.48).

Ceramic Jar GSP201.2 embedded within Concretion GSP201

GSP201.2 consists of an incomplete ceramic vessel, embedded within a concretion (Plate 12, Fig. 3.49). The ceramic vessel is in two parts, one part buried within the concretion, the other being the detached base of the vessel. Both pieces have been partially covered with concretions. The ceramic appears to have been manufactured from coarse grained clay and salt-glazed. The concretion retains a concave 'cast' of the gun GSP201. Although the investigation and stabilization of the ceramic jar would benefit from the removal of the concretion, the aim of the conservation process was to produce a shipwreck object, as a ceramic vessel rising out of a concretion. Some of the concretion and calcium carbonate deposits around the ceramic were therefore removed to reveal the extent of the ceramic surface. This will provide information that can be used to describe its typology, date and location of manufacture.

Mechanical cleaning took place using air abrasive, drills and brushing prior to desalination taking place. This revealed a visually impressive object that will become a stable museum exhibit, alongside the associated copper alloy bowl and concretion GSP201.1.

Preparation of finds archive

On completion of desalination, stabilization and drying treatments, all conserved objects will be transferred for archival storage to Southend Museums (Museum of London Archaeological Archive, 2013). Standard packaging for metal finds has been the use of Plastazote cut-outs within Stewart boxes desiccated with silica gel (Fig. 3.50a). Where necessary, this packaging has been adapted to improve the interpretative potential of the conserved object. For example, with GSP200 (Fig. 51b) the resin casts were packaged in orientation to one another as they appeared within the concretion, leaving gaps where there is a loss in information. This was devised from two colours of Plastazote in black for the flat view and blue for the profile view links. The flat view areas were covered with a layer of soft Tyvek (non-

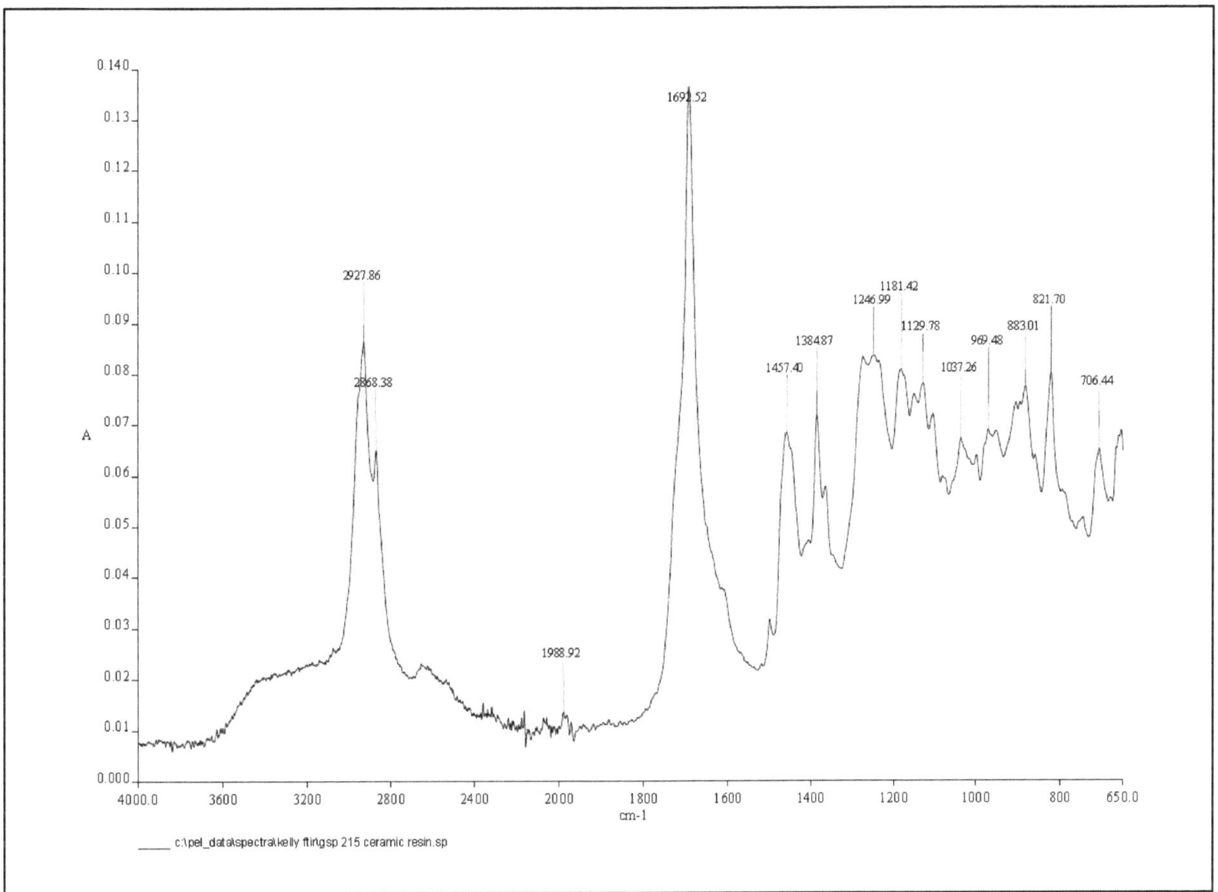

Figure 3.48 *Results of Fourier transform infra-red (FTIR) analysis on pine resin material located on inside of Spanish olive jar fragment (GSP215)*

woven high-density polyethylene fibre) then a custom box was made from Correx (propylene/ ethylene copolymer board).

Conserving the Palimpsest

Perhaps the least archaeologically significant objects to be conserved as part of the Gresham Ship Project are a fragment of textile GSP201.3 (Fig. 3.51) and a concretion containing a modern steel cable GSP26 (Fig. 3.18). However, these objects clearly reveal the dynamic nature of the maritime environment in which the ship wreck materials were preserved. This was not a stable depositional environment that encapsulated a Tudor world, but a dynamic interactive environment in contact with present throughout its time on the seabed. These links in the chain connect the past with the present and enable us to understand the conserved objects of the *Gresham Ship* as a construction of the archaeological process. The results of the conservation process, therefore, have meaning within the constraints of this process, but may be less meaningful outside of this context. This acknowledges the local, personal and unpredictable processes that are evident in the conservation of archaeological maritime material.

The textile fragment was found in association with one of the ship's guns GSP201, preserved within the surrounding concretion material. This object is not from the 16th century ship, but is a 20th-century textile, which is likely to have become incorporated into the wreck via movement of waters and sediment in the channel or perhaps during the channel clearance operation. As such, this modern object has significance as an example of the interaction between the wreck and its surroundings during submersion.

The fragment of synthetic textile has a felted structure and had an extremely regular grid structure pressed onto it, suggesting that it was created by industrial manufacturing methods. The object was a folded fragment of a larger textile (Fig. 3.51). The fibres were impregnated with ferrous corrosion products. The textile was cleaned and desalinated (Jakes and Mitchell, 1992; Landi, 1998; Peacock, 1990), but not all of the deposits, dirt and concretion remains have been removed, as they provide a narrative for the object and are evidence of its association with the wreck and its excavation.

Experimental attempts to air-dry sample fibres resulted in fragile and brittle results. Therefore, the textile was freeze-dried, without prior impregnation at -20°C for three days. After drying the object was more stable, but was still very fragile and difficult to handle without causing damage. The object has been packaged in a 'book' made of nylon net on Correx frames lined on one side with a layer of Plastazote covered in non-dyed calico (cotton textile containing no plasticizers). This packaging allows for both sides of the object to be seen, as the book can be turned over and opened from either side, without directly handling the object.

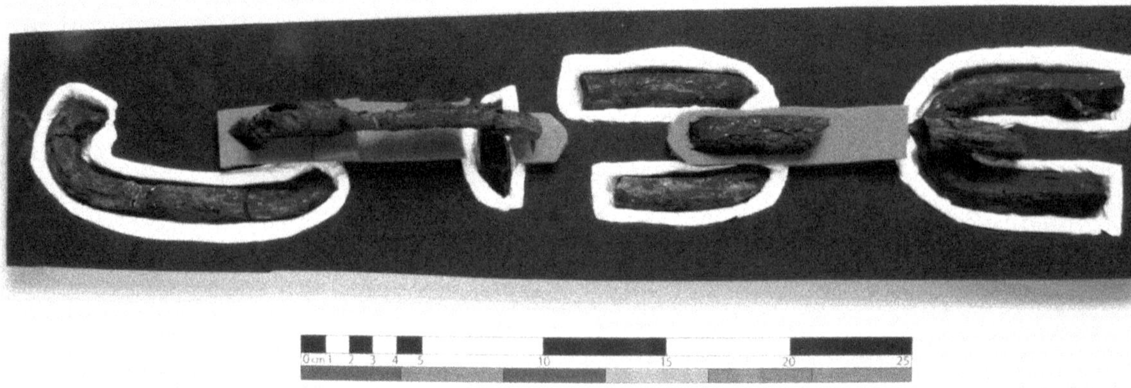

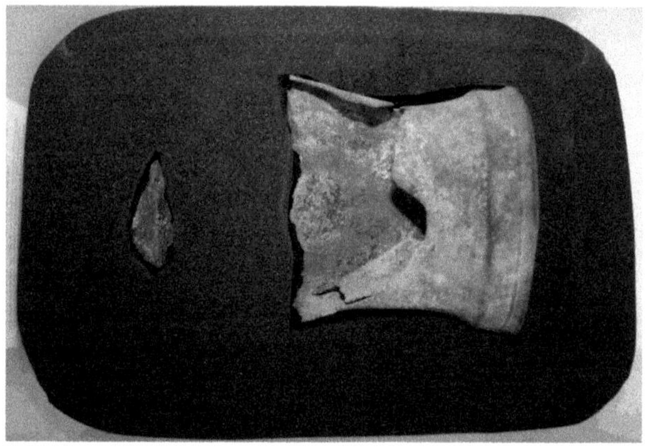

Figure 3.50 Objects packed for archival storage at Southend Museums Service: above resin casts of medium weight chain GSP200; right GSP36 and GSP208 packed for archive deposition (photos Gresham Ship Project)

The presence of recent material, such as the 20th-century fragment of textile GSP201.3 and a concretion containing a modern steel cable GSP26, convey the connections that exist between our time and the moments in time that lead back to the sinking of the ship and the lives of those on board. These objects provide direct tangible links that lead us from the current 21st century excavation and conservation of the finds assemblage to the past of the wreck site. The presence of 19th-century objects, such as the bottle fragment GSP58, provides a further link in the chain that connects us to the sinking of the ship (Fig. 3.52). The sinking is revealed in the chaotic melange of objects thrown together during the wrecking of the ship, preserved as concretions. The copper alloy bowl /concretion (GSP201.1), the ceramic jar/concretion GSP201.2 and the shoe /concretion GSP205 all bear witness to the associations between objects created by the sinking of the ship. Groups of objects used on board ship, such as the tableware, for example a fine silver spoon GSP20.1 and decorative pewter salt holders GSP86.1, 14 connect us with late 16th century life on board this merchant ship. The functional foot wear, well-worn and repaired, (GSP205, GSP206) connect us directly to the crew, while the large over the knee boot GSP80 was perhaps hurriedly kicked off as its owner struggled to leave the sinking ship. The laboriously loaded metal cargo of bar iron and ingots represent the preparations and purpose of the ship's ill-fated final voyage and the guns the dangers of such trading ventures. The ship timbers themselves, with their tool marks and joinery, take us back to the building of the ship and then to the felling of oak trees in the east of England in about 1574 AD (Volume I, Chapter 3.5; Auer and Firth, 2007).

The conservation treatment records contain detailed information about the conservation process and its motivations. These individual records provide a network of individual discoveries during the conservation of the assemblage. Some reveal the excitement of the discovery of a significant object and the challenges of revealing and stabilizing it as a conservation object. Other records reveal more the frustration of similar processes revealing no information that can add to our understanding of the past of the ship and its crew. Critically the documentation of the conservation treatment reveals the latest phases in the lives of the shipwreck artefacts, as archaeological artefacts subject to the processes of heritage conservation. The individual decisions that create the conserved objects are legible as a conservator's signature on each conserved object. These objects are therefore a contemporary creation that have been formed through a network of individual, institutional disciplinary nudges that produce a unique result for each of these real time experiments. When considering these objects as objects of research and objects of communication, the fading contemporaneity of this conservation product should not be misunderstood as anything other than a representation of its time and place, in which the past has been renewed in our time.

The attempt to represent these objects specifically as conserved shipwreck objects will hopefully prevent the temptation for others to present these objects as things from the past that are disconnected with all the moments in time that link our present to how we see our Tudor past.

Conclusion

The changes in society, reflected through an understanding of the technical, cultural, religious, economic and political aspects of the past, can be revealed through the study of wrecked vessels and their accompanying artefact assemblages (Muckelroy, 1998, 24). A ship is a self-sufficient entity, containing everything necessary for its purpose and for the survival, health and recreation of its (usually) male crew (Martin,

1997, 1). The people, practices, techniques, equipment and knowledge that were assembled to sail the ship are symptomatic of the prevailing culture at that time. This can be revealed by examining the objects that are recovered by the archaeological process in the resultant artefact assemblages (Murphy, 1983, 70; Redknap, 1985, 1997; Adams, 2001, 294; Gibbins and Adams, 2000).

The conservation of the Gresham Ship Finds Assemblage successfully stabilized the artefacts and also extracted information for further archaeological research. The vestigial nature of the initial Gresham Ship assemblage (not untypical of rescue archaeology) limited our ability to interpret the lives of those on board or add to our understanding of the Tudor world. To compensate, considerable effort was therefore dedicated to processing the concretions, in a way that may not be the case for more complete shipwreck assemblages.

The absence of objects seen just as voids within the altered form of a concretion, provides an exciting opportunity to use the specialist skills of conservation to retain elements of an archaeological object and present it as a representation of the object as it was at the time of the shipwreck. The approach to the conservation of Gresham Ship assemblage has been of particular importance in demonstrating the possibilities of the conservation process. Significant finds, such as powder chamber GSP202.1, pewter salt holders GSP86.1 and GSP14, spoons GSP86.7 and GSP20.1, copper alloy bowl GSP201.1, pike shaft GSP204, shoes and part of an over the knee boot GSP80, have now been documented, processed, cleaned and conserved for archive deposition. This provides a long-term legacy for future research of the conserved finds. It should therefore be remembered that the conservation process requires arbitrary choices to be made about an object's care, in order to distinguish what physical states are acceptable and what states require intervention. Archaeological objects in the conservation process are not neutral objective truths, but the subject of complex interactions and changing meaning. Conservation is therefore a vehicle to convey and create meaning and value in conserved objects that bear the personality of the conservator. Far from being a neutral process conducted outside of history, conservation is as much a part of the on-going biography of the object as other interventions which have created a trajectory for the object through time (Sully, 2014). In the search for truth and authenticity, conservators may strip away the obscuring layers of history in order to reveal or retain the 'true nature' of the object. Here there is a risk of cutting the threads that link objects with the present which may reduce or limit the stories that can be told about them (Denslagen, 2003, 99). Since every bit of damage and every accretion form part of the evidence of an object's history, their removal can only be justified by making the telling of certain stories easier (Ashley-Smith, 2009, 18).

The editing tools used in this process can be skewed by the artifice of the apparent timelessness of archaeological objects, immutably located in a past state. The work of a practicing conservator can be seen as the latest action in a series of human interactions that affect the physical state of the object. The multiple interactions relate to the creation and use, ownership and exchange of objects and reflect the shifting meaning that objects have within these frames of understanding (Gell, 1998; Avrami, 2009, 183).

The option of conserving an object as a shipwreck artefact as close to its found state as practicable is an important conceptual step towards the aim of retaining as many possibilities for an object's conservation and interpretation in the present and in the future. This Finds Study Programme highlighted the gap between a conceptual idea and the hard realities of dealing with maritime archaeological objects, while suggesting practical approaches to address that challenge.

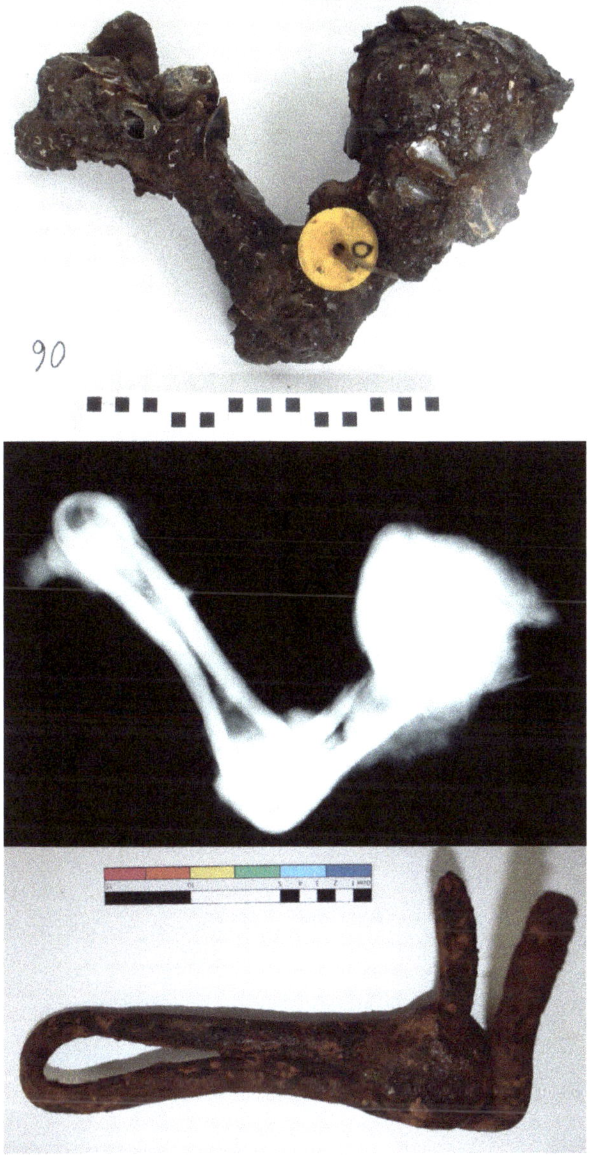

Figure 3.4 (above) Type 1 Concretion: GSP90 contained a well-preserved iron chain link (above) visible in the X-radiograph (centre) and revealed after conservation as part of well-preserved iron chain link GSP90.1 (bottom) (photos Gresham Ship Project)

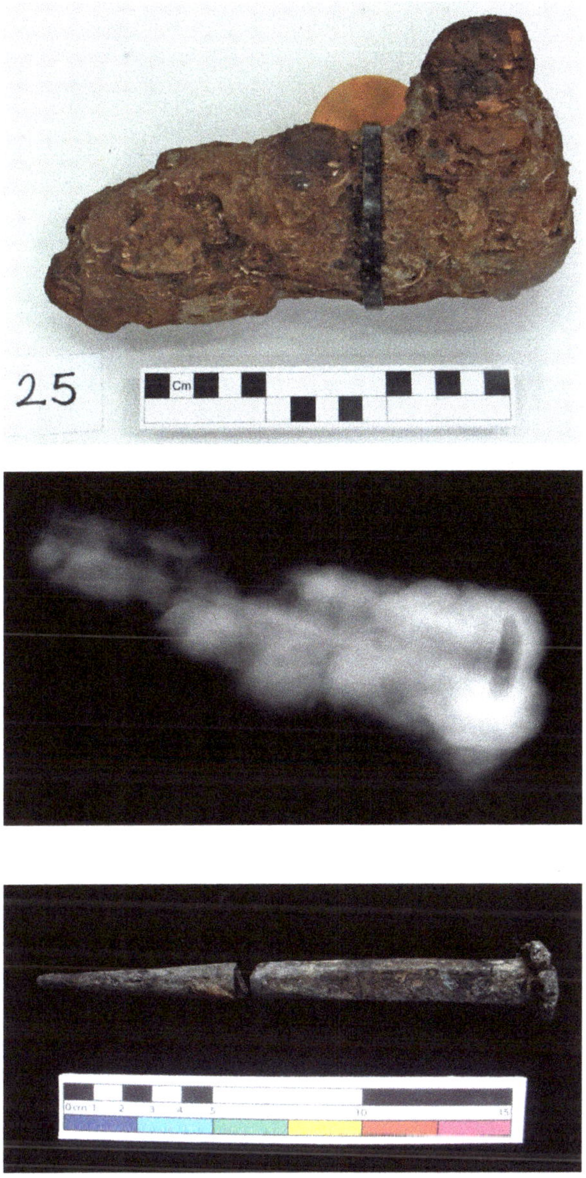

Figure 3.5 (above) Type 2 Concretion: GSP25 contained the void of completely corroded nail (above), as seen in X-radiograph (centre); the completed Polyester resin cast is shown (below), revealing a nail or spike (photos Gresham Ship Project)

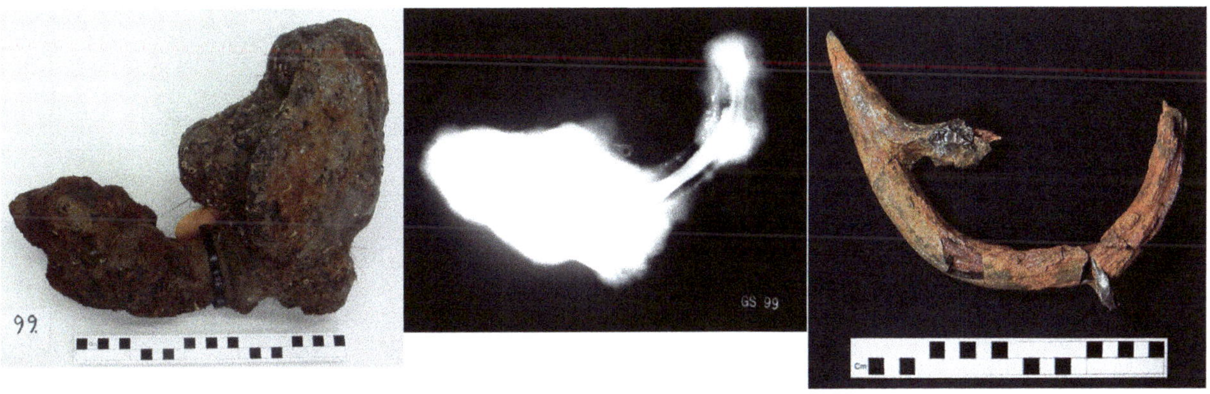

Figure 3.6 (above) Type 3 Concretion: GSP99, containing a hook with solid core (left) and corroded outer edge seen in X-radiograph (centre) and as a composite object partially cast in Polyester resin following conservation (right) (photos Gresham Ship Project)

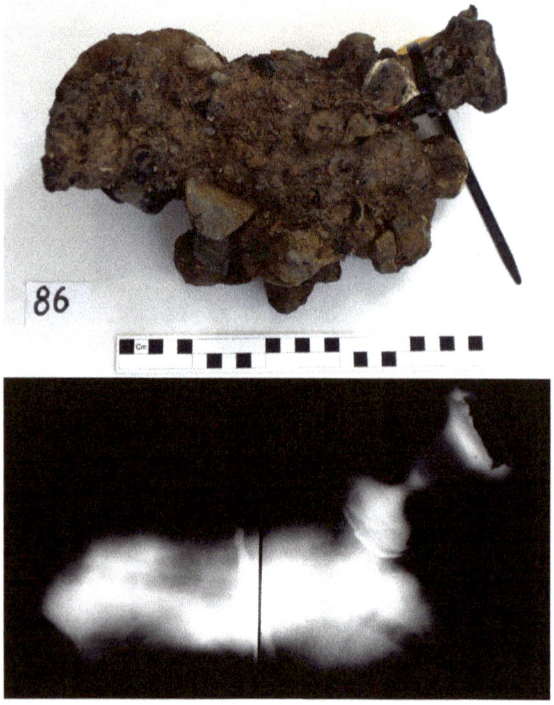

Figure 3.7 (above) The GSP86 concretion (above) contained well-preserved tin-alloy objects (Type 1) and also voids of corroded wrought iron bar and nails (Type 3), shown in the lower image. NB: the black line in the centre of X-radiograph image represents division between two plates (photos Gresham Ship Project)

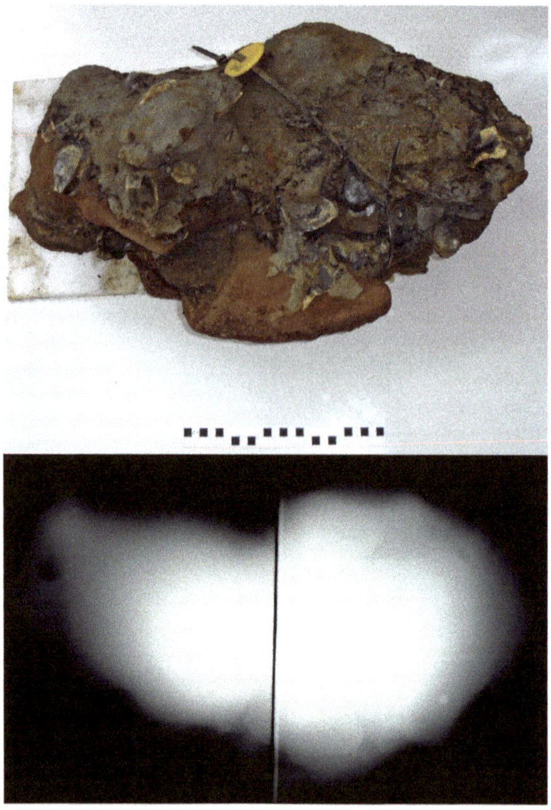

Figure 3.8 Type 4 Concretion: GSP202 containing an object of indistinguishable form and preservation. NB. Black line in the centre of X-radiograph image represents the division between two plates (photos Gresham Ship Project)

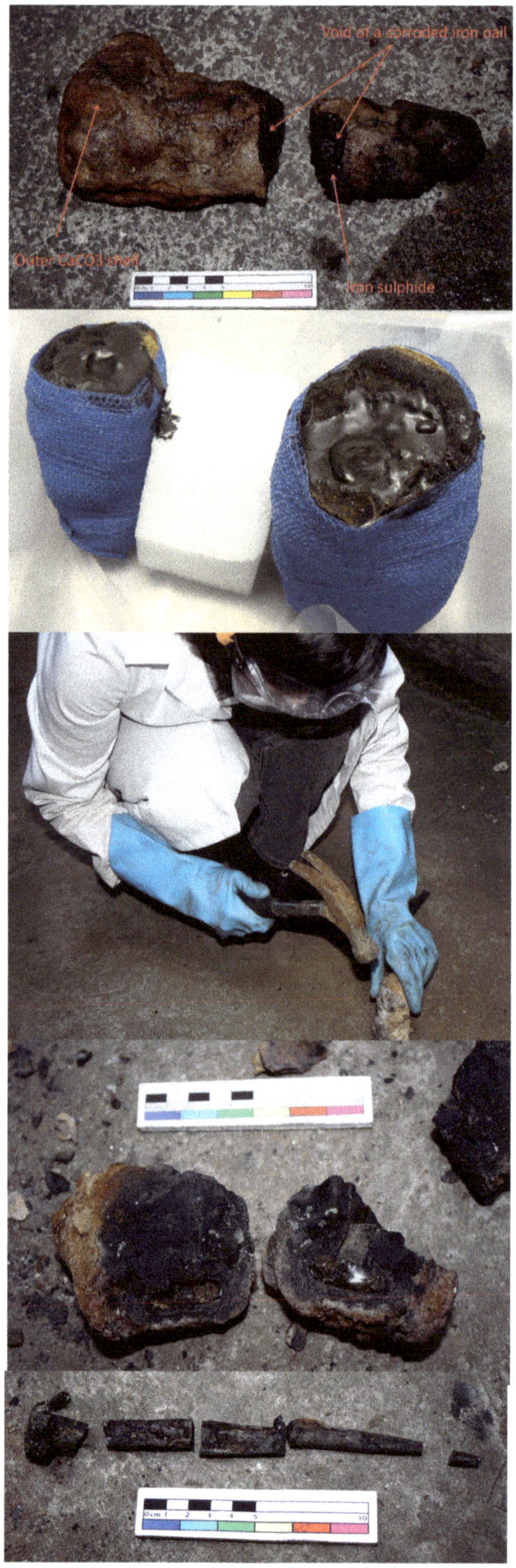

Figure 3.12 (above) Concretion GSP25: (top) initial fracturing; (upper centre) casting the void cavity with Polyester resin and iron weighting powder; (centre) removing the concretion from the resin cast; (lower centre and bottom) retrieving resin cast fragments from concretion (photos Gresham Ship Project)

Figure 3.13 Concretion GS 99: broken into segments (above); excavation of polyester cast (below). See also Figs 3.6 (lower image) and 3.9 (photos Gresham Ship Project)

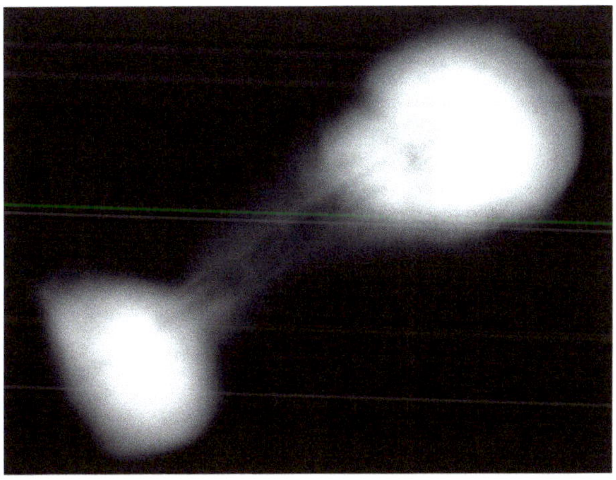

Figure 3.14 Concretion GSP203 (above); X-radiograph of Concretion GSP203 (below) (photos Gresham Ship Project)

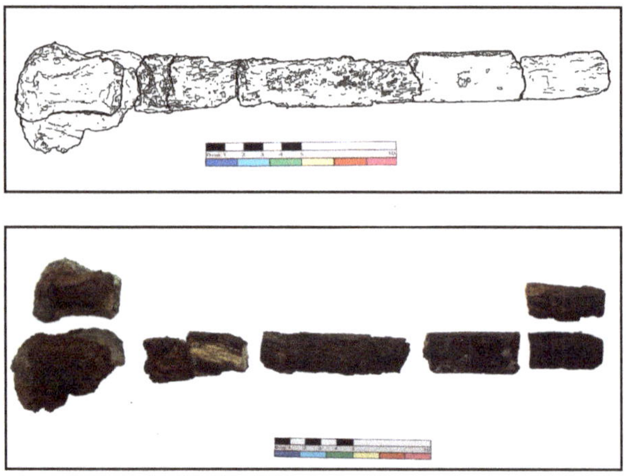

Figure 3.15 Concretion GSP203: reconstruction of metal bolt fragments (photo and drawing Gresham Ship Project)

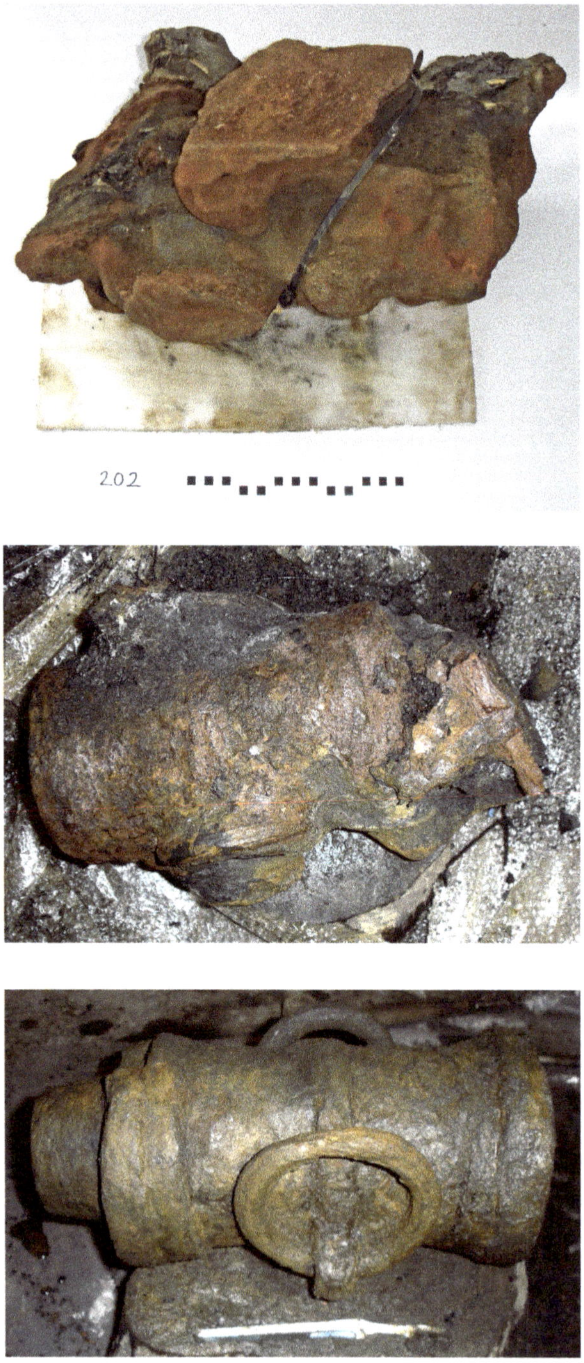

Figure 3.17 (above) Powder chamber GSP202.1 during desalination (photos Gresham Ship Project)

Figure 3.16 (right) Concretion GSP202 before treatment (above), during treatment (centre) with hammer and with powder chamber (GSP202.1) revealed after concretion removal (bottom). For the hammer and shaft found in the same concretion (GSP202.2), see Fig. 3.36 (photos Gresham Ship Project)

Figure 3.19 Salt holders before conservation: above and centre GSP86.1;below GSP14. See also Figs 3.7 lower image, 3.10 and 3.21 (photos Gresham Ship Project)

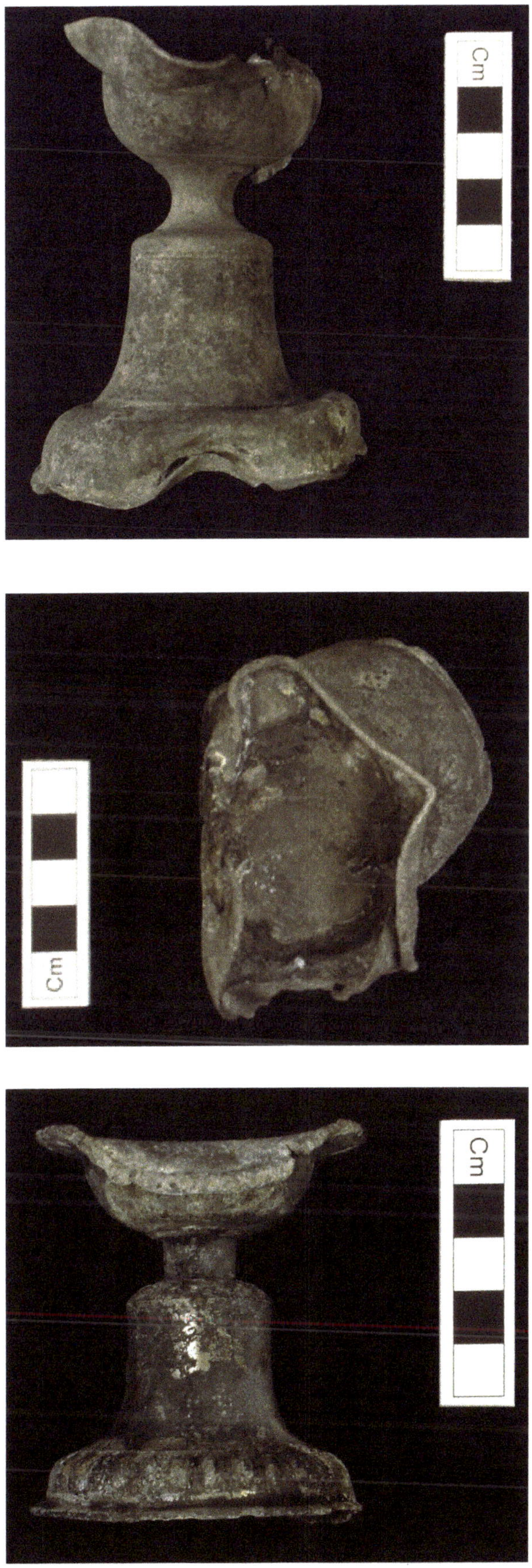

Figure 3.21 Salt holders after conservation: above and centre: GSP86.1; below: GSP14. See also Figs 3.10 and 3.19 (lower image) (photos Gresham Ship Project)

Figure 3.25 Spoon GSP20.1 (top): as removed from Concretion GSP20 (upper centre; spoon with ceramic rim fragment from concretion GSP20; (lower centre) spoon and ceramic rim fragment after cleaning; (bottom) spoon and ceramic rim fragment after cleaning (photos Gresham Ship Project)

Figure 3.28 Copper alloy and concretion GSP201.1 showing riveted areas on base (above) and rim (below) revealed during mechanical cleaning (photos Gresham Ship Project)

Figure 3.29 *Copper-alloy bowl and concretion GSP201.1 after conservation: chemical cleaning with 30% w/v citric acid in de-ionized water applied by a Laponite RD gel poultice and EDTA disodium salt (photos Gresham Ship Project)*

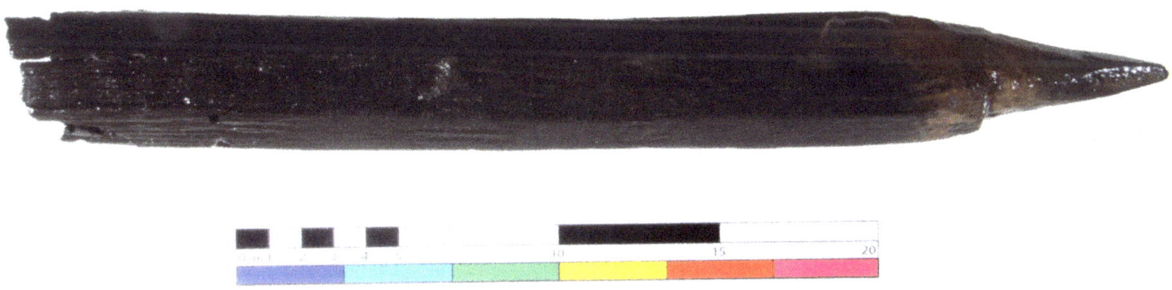

Figure 3.33 *Pike shaft GSP204 (photo Gresham Ship Project)*

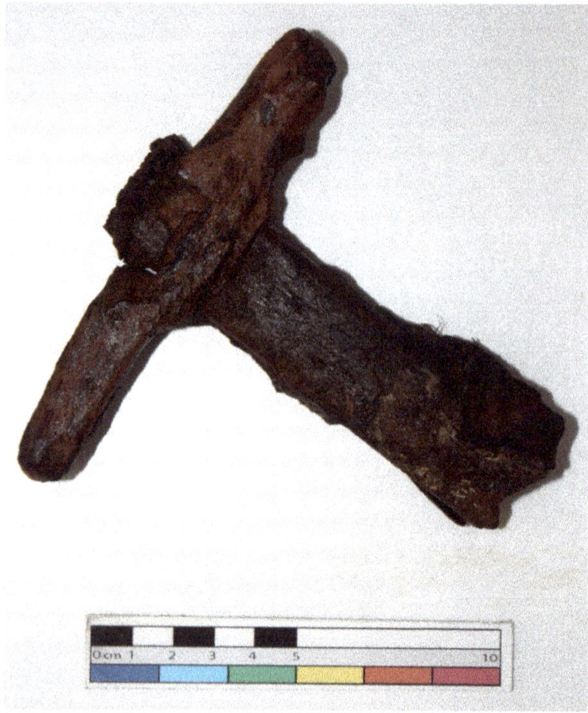

Figure 3.35 Wood and ferrous metal hammer GSP202.2: top left as part of concretion GSP202; centre left and top right as a detached object before treatment; bottom left and right during treatment (photos Gresham Ship Project)

Figure 3.37 Leather footwear as found in 2004: top left over the knee boot GSP80; centre left shoe concretion GSP88; bottom left shoe concretion GSP55; top right shoe sole GSP35; centre right shoe sole GSP205; bottom right shoe sole GSP206. (photos Gresham Ship Project)

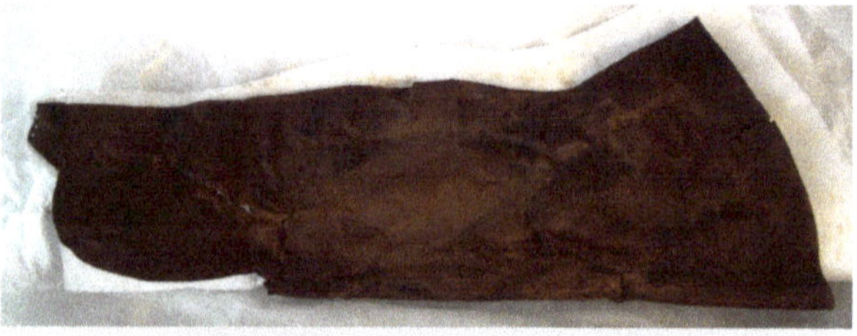

Figure 3.40 left and above: leather boot GSP80 after treatment (photo Gresham Ship Project)

Figure 3.41 Leather shoe GSP88 above before treatment; below during treatment (photos Gresham Ship Project)

Figure 3.42 Leather shoe GSP35: above before conservation; below during conservation (photos Gresham Ship Project)

Figure 3.43 Leather shoe GSP55 after conservation (photo Gresham Ship Project)

Figure 3.44 Leather shoe GSP205 showing wood and iron pike head attached (photo Gresham Ship Project)

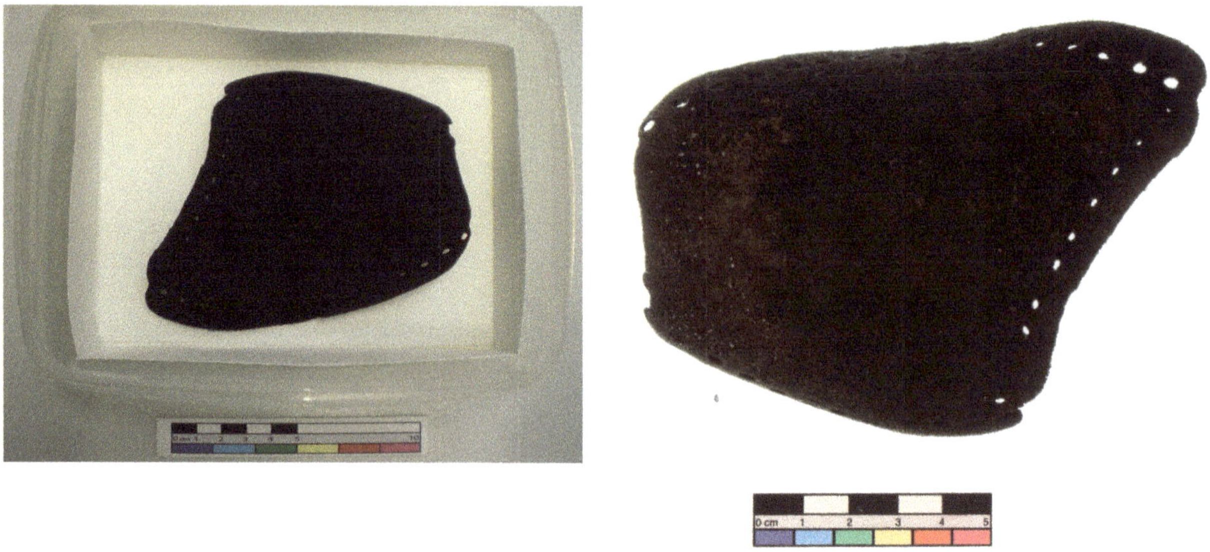

Figure 3.45 *Leather shoe GSP206: left before treatment (116 x 91 x 25 mm); right after treatment (113 x 87 x 26 mm) (photos Gresham Ship Project)*

Figure 3.49 *Ceramic jar and concretion GSP201.2: left above and below before cleaning; right during desalination (photos Gresham Ship Project)*

Chapter 4: The Cargo

by Thomas Birch, Michael F. Charlton, Lynn Biggs, Zofia A. Stos-Gale and Marcos Martinón-Torres[1]

The Gresham ship was carrying a substantial cargo of different metals when she sank, representing a considerable financial loss. Reports record that some of that material recovered in 1846 included 2700 tin ingots, lead pigs and iron bars (British Archaeological Association, 1847, 361). The excavations in 2003–4 found a further three stamped lead ingots, five tin ingots and 42 iron bars from the hull area of the ship. The size of the original cargo remains unknown, though the estimated capacity of the ship, perhaps some 150 tons, provides an indication. In this chapter, the assemblage of lead and tin ingots and the bar iron will be described, the production processes established and the innovative analytical evidence for the provenance of the metals discussed. The research presented here suggests that lead ingots were from Derbyshire, the tin from Cornwall, but the bar iron seems not to be British, but from the Rhine area. Study of such a varied consignment presents an opportunity to learn more about trade in the late 16th and early 17th centuries, setting the cargo in a wider socio-economic context, and leads to suggestions as to a range of possible intended destinations of the vessel before her unfortunate demise.

The technology and provenance of the iron, tin and lead

Methodology

Lead and tin

The iron, lead and tin artefacts were studied and recorded in line with standard archaeological practice before any destructive sampling took place. The lead and tin ingots were transported to UCL for controlled sampling. Two of the lead ingots were sampled twice each using a hack saw, with one sample from the top and another from the bottom. One sample was removed from each of the four available tin ingots. All samples were prepared as standard metallographic blocks at the UCL Wolfson Archaeological Science Laboratories. In an attempt to avoid the grinding grit being stuck on these soft metals, melted paraffin wax was brushed over SiC paper prior to grinding (up to 2500 grade). Lead samples were subsequently polished down to a 0.25µm finish, whereas the polishing of tin was stopped at 1µm. These specimens required no etching and were studied by optical and electron microscopy before undergoing chemical analysis.

The samples were inspected and analyzed using a JEOL 8100 electron microprobe (EPMA) with a wavelength dispersive spectrometer (WDS) at UCL. The same polished blocks were then submitted for further analyses by neutron activation (NAA) and stable lead isotope analysis (LIA) at the Curt-Engelhorn-Zentrum Archäometrie in Mannheim, Germany.

Trace element concentrations were determined by NAA following the protocol described by Kuleff and Pernicka (1995). LIA was performed by multiple-collector inductively-coupled plasma mass spectrometer (MC-ICP-MS). The sample was dissolved in dilute HNO_3 and lead was separated with ion chromatography. The isotope ratios of lead were corrected for mass discrimination by addition of Tl. A value of ^{205}Tl / ^{203}Tl = 2.3871 was used and an exponential relationship assumed for the fractionation within the instrument. ^{204}Pb was corrected for the isobaric interference with ^{204}Hg by measuring ^{202}Hg and using a ^{204}Hg / ^{202}Hg ratio of 0.2293. The in-run precision of the reported lead isotope measurements was in the range of 0.002 to 0.01% (2σ) depending on the ratio considered.

Iron

Fifteen of the recovered iron bars were sampled at the English Heritage Ancient Monuments Laboratory at Fort Cumberland, Portsmouth. One sample was obtained from each bar using a rotary angle grinder with a silicon carbide abrasive disc. Two samples were removed from Bar 78, but none was taken from Bar 39. Each section was then sub-sampled using a tile cutter to remove the areas potentially most affected by the annealing effect of the angle grinder and the specimens were prepared as standard polished metallographic blocks and analyzed at the UCL Archaeological Science Laboratories.

The composition of the metal phase and slag inclusions (SI) was determined using an Oxford Instruments energy dispersive spectrometer (EDS) coupled to a JEOL JXA 8600 superprobe (EPMA-EDS), operating at 20kV with a processing time of 5 and ≈40% deadtime. Further analysis for minor elements in the metal phase and for the glassy phase of SIs was undertaken using a wavelength dispersive spectrometer (WDS) on the same instrument (EPMA-WDS), calibrated using pure elements or

1 M. Martinón-Torres designed and co-ordinated the study, while T. Birch conducted the analyses of the iron bars and related bibliographic research. M. Charlton undertook the iron provenance analysis based on slag inclusion data, L. Biggs carried out metallographic and SEM-EDS analyses of the lead and tin ingots and Z. Stos-Gale interpreted the lead isotopic data. Further bibliographic research was conducted by M. Martinón-Torres, T. Birch and M. Charlton, who wrote the main draft of the manuscript, while all named authors contributed to the writing of the text and approved the final version.

simple compounds. Precision and accuracy were monitored through the analysis of US Geological Survey basalts (BIR-1G, BHVO-2G and BCR-2G) and a British Chemical mild steel standard (BCS. no. 272; SS no. 12). Oxygen was not measured, but was calculated based on stoichiometry.

A minimum of 30 SIs were analyzed using EPMA-EDS area scans at different sites of interest for each sample. Only inclusions greater than 15μm diameter were analyzed in order to reduce localized concentration effects (Dillmann and L'Héritier, 2007, 1815). Each SI was recorded for its location, shape, size, number and type of phases present, so that their information could be combined with observations made through optical microscopy.

In order to avoid the effects of the etchant on the chemistry of the samples, metallographic examination was carried out after EPMA analyses. The samples were etched with a 2% nital solution and examined under an optical microscope for further assessment of texture, carbon content and technological history of the metal.

The lead ingots

Three lead ingots examined were all 'boat'-shaped, with a flat, oval upper surface and a rounded base (Plate 13, Fig. 4.1). Their dimensions ranged from 0.57–0.60 m in length, 190–200 mm in width and 80–100 mm in depth, with recorded weights between 57–60 kg. While their state of preservation is generally very good, much of the surface was covered by white marine encrustations. When this was cleaned mechanically in one of the ingots, multiple impressions of the same circular stamp on the upper surface were revealed (Plate 13, Fig. 4.2). Unfortunately it has not as yet proved possible to identify any parallels for this mark. A deep wedge-like cut is present at one end of this ingot, whose shape suggests it could have been made with an axe (Plate 13, Fig. 4.1).

Boat-shaped lead ingots are known from other post-medieval shipwrecks. Martin (1972, 66–7) noted their presence in the Spanish Armada ships, which carried similar shaped and sized ingots for making into bullets. Lynn Willies (pers. comm.) reports similar – if slightly heavier – examples from a 16th-century wreck found off the Isles of Scilly and tentatively identified as the *San Bartholome*. These ingots are also stamped, but the motif differs those reported here. Further stamped boat-shaped lead ingots have been recovered from vessels associated with Martín Padilla's fleet, wrecked off the coast of Galicia in Spain in 1596, though these weigh up to 120 kg (San Claudio Santa Cruz, 2008). A woodcut in Georgius Agricola's famous metallurgical treatise *De re metallica*, first published in 1556, shows boat-shaped ingots with stamps on their upper surfaces (Hoover and Hoover, 1950, 499). In this source, the lead ingots were to be employed as a reagent in silver extraction by liquation.

Thus it can be shown that this ingot form was a widespread standard shape for storing and trading lead, not necessarily related to either specific makers or specialized uses. In an overview of the typological evolution of lead ingots from shipwrecks dated from the 17th to the 19th century, Martin (2005) notes a trend from the earlier, less standardized, boat-shaped ingots to the parallel-sided and more regular sizes of the later periods. The numerous lead ingots recovered from the Oranjemund shipwreck, a European vessel that sank off Namibia in the 1530s (Chirikure *et al.*, 2010b), are generally more irregular in shape, although many are stamped.

Price *et al.* (1980) show the ease with which boat-shaped ingots could be slung with a rope. Their lifting could be further facilitated by cutting grooves like the one visible in our ingot. Similar grooves were found on ingots from the 1627 *Campen* with a small portion of rope still preserved in one of the cuts (Whiting and Willies, 1985). In any case, it would be relatively easy to cut the soft lead ingots into smaller chunks using an iron axe if needed, as demonstrated by the ingot fragments recovered at the Armada wreck Gran Grifon (Martin, 1972).

Chemical analyses of the Gresham ship's ingots all showed high purity lead, with copper and bismuth being the only elements showing up consistently at trace levels in EPMA area analyses (Tables 4.1 and 4.2). No significant compositional differences were found between the two samples analyzed from each ingot (top

UCL No.	Lab no.	Cu	Fe	Co	Ni	Zn	As	Se	Ag	Sn	Sb	Au
Lead L2-1 (top)	MA-115680	380	100	0.14	37	<14	26.4	3.6	66.7	<200	444	1.81
Lead L2-2 (base)	MA-115681	390	70	0.22	17	<13	10.4	0.57	50.8	84	121	0.15
Lead L3-1 (top)	MA-115682	220	<40	<0.2	20	<6	2	0.68	59.1	<41	54	1.94
Tin T2	MA-115683	580	170	11.5	34	6	377	<6	146	893000	21	1520
Tin T3	MA-115684	760	320	10.1	<81	6	806	<6	3.3	998000	36.2	0.65
Tin T4	MA-115685	580	240	4.5	36	3.5	397	<6	7.7	972000	17.5	0.47

Table 4.1 Major and trace elements in the lead and tin ingot samples analyzed by NAA. All values expressed as parts per million (ppm). The analytical uncertainty is in the order of 2% for As, S, Ag, Sn, and Au, in the order of 10% for Cu and in the order of 20% for the remaining elements. '<' indicates an upper limit of detection in the sample. Tellurium was sought, but was below the detection limit of about 20 ppm. Note that lead cannot be determined with this method

UCL No.	Mn	Ag	Zn	Ni	Sb	As	Fe	Pb	Cr	Cu	S	Ti	Cl	Si	Bi	Sn	Cd	Co
Lead L2-1 (top)	0.003	0.009	0.001	0.002	0.006	0.004	0.000	99.645	0.000	0.014	0.208	0.003	0.046	0.005	0.041	0.000	0.012	0.001
Lead L2-2 (base)	0.002	0.006	0.000	0.001	0.000	0.003	0.000	99.767	0.000	0.034	0.001	0.002	0.079	0.007	0.086	0.000	0.010	0.001
Lead L3-1 (top)	0.004	0.007	0.000	0.001	0.000	0.004	0.000	99.814	0.000	0.015	0.000	0.002	0.066	0.007	0.070	0.000	0.008	0.001
Lead L3-2 (base)	0.002	0.007	0.000	0.002	0.000	0.004	0.000	99.825	0.000	0.011	0.006	0.001	0.066	0.005	0.062	0.000	0.008	0.001
Tin T2	0.004	0.001	0.008	0.003	0.031	0.000	0.016	0.013	0.001	0.024	0.007	0.000	0.003	0.008	0.018	99.854	0.004	0.003
Tin T3	0.002	0.001	0.008	0.002	0.024	0.024	0.032	0.024	0.003	0.051	0.004	0.000	0.002	0.011	0.011	99.798	0.000	0.001
Tin T4	0.002	0.003	0.015	0.001	0.032	0.000	0.017	0.017	0.001	0.010	0.005	0.000	0.000	0.006	0.007	99.880	0.002	0.000
Tin T5	0.002	0.001	0.005	0.001	0.021	0.001	0.016	0.014	0.005	0.041	0.007	0.000	0.001	0.249	0.007	99.622	0.001	0.001

Table 4.2 Average bulk chemical composition by EPMA-WDS of lead and tin ingot samples analyzed. All values are normalized to 100% and expressed in percentage by weight (wt%). Results are averages of 20 measurements of ~100 by 100 μm per sample in the lead and of 12 measurements per sample in each of the tin ingots. Al and Au were sought, but not found above detection limits

and bottom) or between the two ingots. Metallographic examination showed the presence of a harder phase within the lead matrix, which EPMA spot analyses confirmed as lead sulphide (Plate13, Fig. 4.3). This phase appears either in the form skeletal crystals or as amorphous clusters, often surrounded by post-depositional oxidation; the latter form predominates in the samples taken from the top of the ingots. While the skeletal growths are clearly recrystallizing from the melt upon cooling, the more irregular form is likely to represent residual, partly unreacted galena from the ore. Its higher concentration towards the top of the ingots can be explained as a result of its lower density compared to metallic lead, which would cause them to rise. More rare inclusions high in copper (20–30%) and/or antimony (20–35%) were recorded too, with occasional presence of sulphur and/or nickel. Spot analyses of these by EPMA sometimes showed boron levels of up to 1.5%, likely an enrichment resulting from their deposition in seawater.

The lead isotope compositions of the three samples analyzed (taken from two ingots) are identical within ±0.01% for all independent three lead isotope ratios and show that they were most likely made from lead smelted from the same ores, possibly in the same metallurgical operation (Table 4.3). These signatures were compared with those for geological lead ores in the OXALID database using TestEuclid (Stos-Gale and Gale, 2009). The only sets of ratios for lead ores matching our metal samples are those from the north English lead mines in Derbyshire (mines near Castelton, Hopton, Matlock) and Cheshire (Alderley Edge). Lead isotope data for these ores were published by Rohl (1996, available on the OXALID database: http://oxalid.arch.ox.ac.uk/British/British.html). Figure 4.4 (Plate 4) shows the mirror lead isotope ratios plots of the data for these ores and the lead ingots from the Gresham ship.

The lead and copper ore deposits of Alderley Edge were first exploited in the Bronze Age and there is also evidence of Roman mining. However, after the period of Roman activity there are no further records of mining until the late 17th century (Timberlake and Prag, 2005). On the other hand, the lead mines of Derbyshire were active throughout our period of interest: between 1560 and 1600 lead production increased eleven fold and the trend was further accelerated in the following thirty years, largely thanks to technological developments and a growing export market (Kiernan, 1989, 269; Wood, 1999, 72–3). On the account of this well-documented lead exploitation in 16th-century Derbyshire and a complete consistency of lead isotope ratios measured for the lead ingots from the Gresham ship with those for lead ores from Hopton (Snake Mine), Castleton and Matlock, we can conclude that the ingots originated from lead smelted in the Peak District of Derbyshire.

The tin ingots

Five tin ingots recovered from the Gresham shipwreck were available for study, of which four could be

UCL No.	Lab no.	$^{208}Pb/^{206}Pb$	$^{207}Pb/^{206}Pb$	$^{208}Pb/^{204}Pb$	$^{207}Pb/^{204}Pb$	$^{206}Pb/^{204}Pb$
Lead L2-1 (top)	MA-115680	20.798	0.84628	38.397	15.624	18.462
Lead L2-2 (base)	MA-115681	20.797	0.84624	38.400	15.625	18.464
Lead L3-1 (top)	MA-115682	20.796	0.84594	38.423	15.630	18.476
Tin T2	MA-115683	20.652	0.83975	38.458	15.638	18.622
Tin T3	MA-115684	20.693	0.84175	38.441	15.637	18.577
Tin T4	MA-115685	20.674	0.84042	38.468	15.638	18.607

Table 4.3 Lead isotope ratios for the investigated samples. NB: the last two ratios were calculated from the previous three

sampled for analysis. They are 'strip-like' in appearance (Plate 15, Fig. 4.5), four complete at lengths 0.58–0.60 m, one broken at 0.36 m. The cross-section is trapezoidal, with top sides varying 15–22 mm, bottom sides 11–15 mm and thickness 6–10 mm. On the upper surface there is evidence of shrinkage in the form of a longitudinal depression running across the middle and some possible remnants of casting spurs. The 19th-century report on the wreck site (British Archaeological Association, 1847, 361) notes the presence of a tin ingot with a royal stamp (a rose surmounted by a crown), but no such stamp could be identified in the ingots studied here.

Ingots of similar shapes have been recovered in the Rooiberg area, South Africa, and thought to have been made from local ores in a large tin smelting industry that flourished from the mid-second millennium AD (Chirikure *et al.*, 2010a). Some samples of these were analyzed by NAA (Grant, 1990) and found to be of lesser purity than the ones studied here, however, containing copper and cobalt as the major impurities (see below). In addition, bar-shaped ingots 50 cm long are described in the cargo from the Oranjemund wreck (Chirikure *et al.*, 2010b).

Chemical analyses demonstrate the comparatively high purity of these ingots, with average values of elements such as copper, iron, lead and arsenic below 0.1% (Tables 4.1 and 4.2). Spot analyses in the grain boundaries, however, detected a localized enrichment of arsenic (<40%), iron (<30%) and/or copper (<25%) with traces of nickel and/or cobalt (generally <0.5%). The metallographic investigation also revealed composite clusters of iron-tin crystals towards the bottom surfaces of the ingots (Plate 15, Fig. 4.6). Spot analyses by SEM-EDS and EPMA confirmed the identification of most of these phases as FeSn, sometimes containing a core of Fe_3Sn_2.

Iron-tin alloys are a common impurity formed during tin smelting and they are often referred to as 'hardhead' because of their deleterious effects on the metal properties. Hardhead, however, tends to float towards the top of the metal, given its lower density. In this case, the crystals are concentrated towards the mould surfaces, suggesting that the local enrichment in iron may derive from the material of the mould where the ingots were cast. While cooling in a tin-rich bath, Fe_3Sn_2 crystals would progressively transform into FeSn from the outside in; upon solidification, any Fe_3Sn_2 left would remain as a core surrounded by FeSn within the pure tin matrix, as observed. The slow cooling would also have allowed any copper and arsenic to migrate to the phase boundaries. On balance, it would seem that these ingots were cast from relatively pure tin in iron moulds, possibly after refining or at least remelting rather than straight from a smelting furnace.

The lead isotope ratios measured for these three ingots seem to be also quite similar, indicating that the tin metal was smelted from ores from the same mining district (Table 4.3). These isotope compositions obtained not at all typical for tin ingots when compared to many previously analyzed at Oxford, from the Uluburun and various other shipwrecks in the eastern Mediterranean (Stos-Gale *et al.*, 1998, Clayton *et al.*, 2002) to the 17th-century Dutch shipwreck from Aanloop Molengat (unpublished data from the Isotrace Laboratory). Due to the geochemistry of tin ores, they occur most often in granite rocks associated with grains of uranium and thorium that continuously produce more radiogenic lead isotopes. This process influences the lead isotope compositions of tin ores making them 'radiogenic', that is, with a $^{208}Pb/^{206}Pb$ ratio lower than <2.04. Unusually, this ratio approaches 2.07 in the ingots from the Gresham Ship. The comparison with all the available ore data shows them to be fully consistent (within ±0.1%) with only one ore sample: a malachite from the St. Day Mine in Truro, Cornwall (plotted on Fig. 4.4, Plate 14).

The levels of lead in cassiterites from Cornwall measured by lead isotope dilution are in the order of 10–60 ppm (measurements made by Noel Gale and Robin Clayton, Isotrace Laboratory, Oxford), while in copper and iron ores these concentrations can reach several hundreds or even thousands of ppm. Hence, even if minute amounts of copper minerals were mixed with the tin during the smelting or further processing, they could dominate the lead isotopic signature of the resulting ingots. The small but perhaps significant enrichment of copper in the tin ingots noted above may therefore explain the match between the Gresham tin and the Cornish copper mineral. Alternatively, the distortion of the lead isotopic signature of the tin metal might also be due to contamination with a small amount of local lead minerals, as perhaps indicated by the bulk levels of lead in the tin ingots, in all cases higher than 1000 ppm according to EPMA data – note that Pb is not measured by NAA (Table 4.2).

A comparison was made of the lead isotope ratios measured for T2 and T4 with the 17th-century tin ingots from *Anloop Molengaat* (supposed to have originated from the Erzgebirge) and with the lead isotope data for cassiterites and various ancient tin ingots from Cornwall (Isotrace Laboratory, unpublished data). This comparison is plotted on two mirror diagrams in Fig. 4.7 (Plate 16). On the smaller scale of this plot, the lead isotope ratios of the three tin ingots from the Gresham ship seem identical to the signature of cassiterites and tin ingots from Cornwall. The direct comparison of their lead isotope ratios does not make them identical, but nevertheless they are fully compatible with the mixing lines of tin ores from Cornwall. It is therefore compelling to conclude that these ingots are indeed made of tin from the Cornish ores.

Iron technology and provenance

Although iron bars or bar fragments were scattered around the wreck site of the Gresham ship, the majority were recovered from one area in the hull of the ship. A summary description of the iron bars can be found in Table 4.4. Several exhibit heavy surface corrosion in part or in their entirety, but despite this most have maintained their integrity with their metal substrate largely unaffected. Of the sixteen bars available for study, nine are considered complete. These are folded onto themselves, forming packages around 2m in length.

On initial visual inspection, two bar types were plainly visible. Two bars have a rectangular cross-sections (≈2 x 90 mm), described here as 'flat', whilst the remainder are square (≈4 x 40 mm). Of the square-sectioned bars (Type A), closer examination reveals there to be two sub types; those with one fold towards the middle (Bars 43 and 47) and those with multiple folds. Four of the multi-fold bars show the same three-fold characteristic (Bars 42, 44, 45 and 78), whilst Bar 40 shows two folds only – though the appearance of the terminal of this bar is indicative of a break at a fold. The 'flat' bars are labelled Type B (Plates 17 and 18, Figs 4.8–4.10). The seven iron bar fragments show fibrous and bent terminals, comparable to those in the complete bars, indicating that they had been part of a larger intact iron package. None of the bar fragments were assigned to a 'Type', since it was impossible to determine their original morphology. Bars 22 and 46 appear to be the larger main segments of former bars, whilst the others are considerably smaller, resembling the 'arms' of the complete bars (Plate 18, Figs 4.11–4.12).

Technological origin: two compositional groups of bloomery iron

One of original aims of this study was to identify the process used to manufacture these bars. Since there are so few known examples of bar iron, the Gresham material provides invaluable insights into the production of these items during the period.

There were two iron-making methods in Europe at this time. The first method is the traditional direct or bloomery process, where ores rich in iron oxides are reduced to form a metallic mass, a 'bloom' in the solid state. The second method, which was gradually being adopted throughout Europe, was the indirect process.

TYPE	BAR	FOLDS	DIMENSIONS				WEIGHT (kg)	FEATURES	CORROSION
			cross-section (mm)		length (cm)	unfolded (cm)			
A	40	2	square	45x40	203	512	56.5	ham.	parts heavily
A	42	3	square	35x35	184	565	55.0	ham.	parts heavily
A	43	1	square	40x30	188	372	42.0	ham.	parts heavily
A	44	1	square	40x40	193	569	70.0	ham.	parts heavily
A	45	3	square	45x40	201	563	65.0	ham.	parts heavily
A	47	1	square	45x40	190	381	40.5	ham.	minor
A	78	3	square	40x35	183	554	52.0	ham.	minor
B	18	2	flat	90x20	197	425	60.0	rolled?	minor
B	76	2	flat	90x25	188	415	65.0	rolled?	minor
frag.	17	-	square	30x30	160	160	9.0	ham.	minor
frag.	22	1	square	45x40	191	191	60.0	ham.	heavily
frag.	30	-	square	30x30	98	100	3.0	ham.	minor
frag.	38	-	square	40x40	88	87	3.5	ham.	heavily
frag.	39	-	square	35x25	104	104	7.0	ham.	heavily
frag.	46	1	square	40x40	181	287	28.0	ham.	parts heavily
frag.	48	-	square	40x30	108	108	5.5	ham.	minor

Table 4.4 Summary description table for the iron bars, including dimensions, observed features and group assignment. Key: frag = fragment; ham = hammered

BAR	TYPE	n=	Na_2O	MgO	Al_2O_3	SiO_2	P_2O_5	SO_3	K_2O	CaO	TiO_2	MnO	FeO	BaO	Total
17	A	30	0.4	1.0	3.1	30.6	0.4	0.3	2.5	2.6	0.1	2.7	56.3	nd	103.1
18	B	40	0.6	1.1	4.0	27.6	bd	nd	2.6	2.8	0.1	3.3	58.0	nd	105.7
22	A	30	0.3	2.5	2.9	30.4	0.5	0.9	2.3	3.7	bd	15.0	41.2	0.4	103.6
30	A	31	0.3	3.3	3.3	41.3	0.0	0.5	3.7	4.9	0.1	18.4	23.4	0.8	102.0
38	A	30	0.8	2.5	7.2	47.3	0.2	0.2	5.6	6.1	0.3	4.4	25.5	nd	103.7
40	A	31	0.7	2.0	6.2	44.0	0.3	0.2	4.4	5.2	0.2	3.9	32.9	nd	102.8
42	A	31	0.3	0.9	2.8	31.7	0.5	0.2	2.4	2.2	0.1	2.4	56.5	nd	101.9
43	A	30	0.3	2.0	5.0	44.6	0.1	0.2	4.3	6.3	0.2	17.5	18.6	0.7	102.1
44	A	30	0.6	1.6	4.7	35.2	0.5	0.2	3.7	4.3	0.2	4.2	44.8	nd	102.3
45	A	31	0.3	0.9	3.9	28.9	0.6	0.2	1.9	3.7	bd	2.2	57.4	nd	102.4
46	A	30	0.7	2.5	7.2	52.0	0.2	bd	6.0	6.4	0.3	3.8	20.8	nd	102.2
47	A	30	0.4	1.2	4.8	29.7	0.6	0.1	3.2	3.6	0.1	3.3	53.1	nd	102.7
48	A	30	0.5	1.8	6.5	37.9	0.2	0.2	4.6	5.4	0.2	3.7	39.0	nd	100.9
76	B	30	0.4	1.1	3.8	33.9	0.6	0.1	3.2	3.1	0.1	3.3	50.4	nd	101.8
78 S1	A	30	0.7	2.1	6.1	50.7	0.3	0.1	4.6	4.8	0.2	5.2	25.2	0.0	102.9
78 S2	A	30	0.5	1.2	3.9	30.3	0.4	0.0	2.6	2.9	0.0	3.9	54.1	0.0	101.2

Table 4.5 Normalized average EPMA-EDS data from analysis of SIs in iron bars. Results expressed as stoichiometric oxides in wt% and the 'Total' column provides average analytical total prior to normalization. Key: bd = below detection limits; nd = not detected; n = number of analyses

This used blast furnaces operating at much higher temperatures to produce liquid, carbon-rich cast iron. The cast iron was subsequently decarburized at a later stage to make it malleable. Though the indirect process required two steps – smelting and decarburization – it also allowed the fruitful exploitation of lower-grade ores. Whatever the initial technological origin, the final shapes of low-carbon iron artefacts were achieved through iterative cycles of heating and hammering, or 'smithing'.

The origins, development and spread of blast furnace technology throughout Europe has been much debated (Awty, 1987; Blick *et al.*, 1991; Crossley, 1984; Gömöri, 1986; Morelli, 1982; Reynolds, 1986; Tholander and Blomgren, 1986; Tylecote, 1986). The earliest use of this technology originates in China, appearing much later in Europe. Three regions have been identified as having the earliest blast furnaces in Europe, dating from the early 13th to the 15th century; they are Lapphyttan (Sweden), Dürstel (Switzerland) and Märkische Sauerland (Germany) (Jockenhövel and Willms, 1997; Magnusson, 1995, 2011a). By the late 16th century, the blast furnace had been introduced and was becoming established throughout most of Europe (King, 2005; Tylecote, 1976; 1987; 1991), but they did not immediately replace all the older high-bloomery furnaces. There is general agreement that the new and old technologies co-existed for some time. In parts of Sweden, bloomery iron production was still widely practised as late as the 18th century (Englund, 2011). There are, of course, regional and chronological differences between different iron producing regions across Europe. The smelting technology is therefore highly relevant when discussing the likely provenance of the Gresham bars and situating them in a wider technological and economic context.

Despite several attempts, no conclusive method has been developed to distinguish iron made by different production processes. However, general observations made about the differences between direct and indirect produced iron can be informative (Buchwald and Wivel, 1998; Gordon, 1997; Hedges and Salter, 1979; Mackenzie and Whiteman, 2006; Rostoker and Dvorak, 1990; Starley, 1999; Tholander, 1989; Williams, 1990). In order to distinguish between processes, these studies have largely focused on the differences of the composition, phases and morphology of slag inclusions (SI). These are generally small but abundant stringers of slag that can be identified within iron objects. SIs are remnants of the original smelting slag – particularly in bloomery iron – though their chemistry can be altered – and new SIs created – during subsequent episodes of smithing, decarburization or refining. Dillmann and L'Héritier (2007) presented a means to distinguish direct and indirect-produced iron based on the relative amounts of phosphorous to other oxides in the SI. Through experimental studies they demonstrated that iron made in a blast furnace will contain SIs richer in phosphorous and poor in alumina and alkali earth oxides (relative to iron), whilst bloomery SIs will show the opposite trend.

The average SI composition for each of the Gresham bars is presented in Table 4.5, listing the major and minor element oxides detected. Table 4.6 shows the average SI composition that was obtained by

BAR	TYPE	n=	Na_2O	MgO	Al_2O_3	SiO_2	P_2O_5	SO_3	ClO	K_2O	CaO	TiO_2	MnO	FeO	CoO	NiO	CuO	ZnO	AsO_2	SnO_2	PbO	Probe Total
17	A	3	0.557	1.252	3.991	38.181	0.362	0.479	0.005	2.498	3.352	0.166	3.034	47.739	0.041	nd	0.027	0.015	0.054	0.006	0.055	101.751
18	B	4	0.741	1.279	5.077	37.197	0.243	0.177	0.004	2.765	4.149	0.138	3.810	47.381	0.038	0.007	0.037	nd	nd	nd	0.053	103.056
22	A	3	0.333	3.382	3.871	38.495	0.186	1.003	0.012	2.534	5.001	0.134	17.336	28.644	0.039	0.016	0.008	0.029	0.052	0.002	0.061	100.981
30	A	3	0.536	3.865	4.713	49.850	0.034	0.400	0.013	3.486	7.139	0.245	18.903	9.969	0.004	0.012	0.014	nd	nd	0.005	0.059	99.171
38	A	3	0.769	2.474	8.831	55.770	0.168	0.222	0.032	3.064	7.036	0.274	5.044	17.690	0.019	0.002	0.016	0.016	nd	0.020	nd	101.406
40	A	3	0.649	2.654	9.150	67.354	0.023	0.029	0.009	2.951	7.708	0.304	4.900	3.213	0.008	0.019	0.029	0.007	0.039	0.013	nd	98.977
42	A	3	0.431	0.767	3.146	32.128	0.503	0.303	0.007	2.118	2.624	0.144	1.913	56.924	0.053	0.006	0.015	0.007	0.062	nd	0.156	101.279
43	A	3	0.300	1.198	3.962	33.646	0.329	0.326	0.007	2.075	4.911	0.146	13.991	38.566	0.018	nd	0.024	0.022	0.073	0.016	0.080	99.628
44	A	3	0.645	2.749	8.081	61.317	0.045	0.152	0.010	2.683	8.076	0.241	6.853	9.603	0.008	0.002	0.008	0.004	nd	0.006	0.008	100.467
45	A	5	0.848	0.360	6.905	33.981	0.785	0.426	0.068	2 815	6.359	0.115	1.420	48.904	0.040	0.012	0.016	0.009	0.068	0.014	0.058	103.099
46	A	3	0.618	2.187	6.265	46.392	0.242	0.070	0.007	3.039	5.718	0.209	4.594	30.107	0.013	0.002	0.005	0.033	0.020	0.010	0.028	99.448
47	A	3	0.512	1.287	5.940	38.672	0.551	0.159	0.005	2.516	4.719	0.122	3.650	44.346	0.048	0.009	0.010	nd	nd	0.007	0.045	102.594
48	A	3	0.642	2.894	10.281	62.007	0.033	0.033	0.010	2.949	8.999	0.358	4.228	6.780	0.016	0.006	0.007	0.008	nd	nd	0.044	99.235
76	B	3	0.546	0.995	4.187	35.214	0.472	0.070	0.009	2.020	3.262	0.146	3.197	49.764	0.045	nd	0.040	0.007	0.064	0.002	0.042	100.043
78 S1	A	3	0.689	1.025	4.796	32.203	0.855	0.408	0.089	1.909	2.950	0.125	3.445	52.648	0.054	0.004	0.021	0.013	nd	0.010	0.057	101.264
78 S2	A	3	0.278	1.523	1.946	31.951	0.414	0.029	0.011	C.736	2.348	0.063	4.386	59.392	0.039	0.002	0.031	0.023	0.016	0.009	0.043	103.206

Table 4.6 *Unnormalized average EPMA-WDS data obtained from analysis of SIs in iron bars. The results are expressed as elements in wt% and the 'Total' column provides the analyticaltotal. Key: nd = not detected; n = number of analyses*

BAR	TYPE	n=	Al	Si	P	S	Cl	Ti	Cr	Mn	Fe	Co	Ni	Cu	Zn	As	Mo	Sn	W	Pb	Total
17	A	5	nd	nd	0.043	0.007	0.008	0.003	0.005	0.016	100.229	0.090	nd	0.079	0.004	0.076	nd	nd	0.017	0.068	100.555
18	B	7	nd	0.001	0.018	0.015	0.008	0.004	0.007	0.025	101.119	0.091	0.006	0.030	0.013	0.035	nd	0.018	0.025	0.080	101.391
22	A	10	nd	nd	0.034	0.062	0.009	0.006	0.005	0.027	98.993	0.106	0.013	0.033	0.013	0.043	nd	0.007	0.026	0.075	99.373
30	A	5	0.005	nd	0.019	0.010	0.008	0.007	0.005	0.026	98.338	0.083	0.007	0.035	0.007	0.039	nd	0.018	0.034	0.051	98.617
38	A	10	0.002	nd	0.048	0.007	0.004	0.006	0.003	0.013	99.880	0.091	0.011	0.030	0.010	0.078	0.015	0.008	0.045	0.053	100.201
40	A	5	nd	nd	0.036	0.009	0.004	0.003	nd	0.043	99.987	0.100	0.013	0.014	0.004	0.183	nd	0.008	0.010	0.079	100.317
42	A	10	nd	nd	0.029	0.010	0.007	0.004	0.007	0.011	100.160	0.092	0.013	0.042	0.019	0.039	0.019	0.013	0.028	0.080	100.556
43	A	5	0.004	nd	0.010	0.007	0.025	0.007	0.002	0.038	99.906	0.094	0.008	0.020	0.015	0.032	nd	0.007	0.019	0.095	100.205
44	A	5	nd	nd	0.028	0.012	0.008	0.003	0.007	0.023	99.514	0.084	nd	0.034	0.010	0.078	nd	0.006	0.013	0.092	99.853
45	A	5	nd	nd	0.018	0.019	0.009	0.007	0.006	0.015	99.032	0.089	0.002	0.018	0.003	0.050	nd	0.002	0.035	0.073	99.326
46	A	10	nd	nd	0.040	0.003	0.008	0.004	0.009	0.012	99.930	0.102	0.007	0.053	0.016	0.034	0.017	0.005	0.018	0.066	100.214
47	A	10	nd	nd	0.027	0.007	0.010	0.006	0.007	0.014	100.576	0.096	0.010	0.037	0.016	0.053	0.002	0.003	0.043	0.091	100.884
48	A	5	nd	nd	nd	0.009	0.011	0.004	0.007	0.009	98.606	0.087	0.006	0.023	0.019	0.122	nd	0.022	0.012	0.047	98.874
76	B	5	nd	nd	0.066	0.002	0.007	0.002	0 003	0.018	98.316	0.086	0.010	0.144	0.012	0.041	nd	0.012	0.030	0.074	98.745
78	A	10	nd	nd	0.069	0.007	0.006	0.008	0.005	0.012	99.842	0.094	0.005	0.115	0.016	0.058	0.014	nd	0.033	0.059	100.243

Table 4.7 *Unnormalized average EPMA data obtained from analysis of metal phase, expressed as elements in wt%. Key: nd =not detected; n = number of analyses*

microprobe analysis, supplementing the major and minor elements with additional trace element concentrations. The metal composition for each bar is provided in Table 4.7.

Based on the method developed by Dillmann and L'Héritier (2007), Fig. 4.13 (Plate 19) shows a plot of the composition of the SIs analysed from the Gresham bars compared to reference data from known direct and indirect produced iron. It is clear that the Gresham bars are made from bloomery iron as the phosphorous contents of SIs are relatively low in all cases. Bars 30 and 43 contain very few SIs containing phosphorous, with none present in any SI in bar 18, whilst less than half of all SIs analysed in the remaining bars detected phosphorous. None of the SIs exceed 1 wt% in their average phosphorous contents.

When the SI data from the Gresham bars was subjected to further multivariate statistical testing, two distinct groups were identified in the bloomery iron used, shown in Fig. 4.14 (Plate 19). Hierarchical cluster analysis (HCA) and principal component analysis (PCA) were performed on normalized sub-compositions (cf. Birch and Martinón-Torres, forthcoming), which showed that Bars 22, 30 and 43 formed Group I, whilst the remainder formed Group II, as is visible in Fig. 4.14. It is important to note from the outset that no correlation was found between the compositional groups (using SI) and the typological groups identified (Types A and B).

Group I is characterized by higher amounts of MnO, BaO and MgO relative to SiO_2. The mean average of MnO for Group I is 17.4 wt%, which is almost five times higher than Group II (3.5 wt%). The increased MnO contents can only be explained by the use of a different ore, or different ore mixture, as MnO is minimal in clay (furnace lining) and fuel ash from charcoal (fuel) combustion. The chemical homogeneity of SIs for each bar sample studied strongly supports the hypothesis that each section represents a raw bloom. These raw blooms are chemically similar, revealing that they come from the same smelting system (either Group I or Group II), with subtle differences between smelts.

In order to produce the Gresham iron bars from the direct process, their construction would have required the welding together of several blooms. Besides the welding lines noticed in the metallographic examination, the welding together of raw blooms is apparent from the SI analyses of two samples from Bar 78. The chemistry of the SIs analysed in the two samples taken from different parts of the same bar overlap, indicating that they derive from the same smelting system, but show subtle differences enough to distinguish two chemical subgroups, consistent with two similar blooms being welded into the same bar (Fig. 4.14, Plate 19).

The chemical analysis of the metal phase in the Gresham bars shows it to be very pure iron. EPMA data showed only traces of cobalt at around 0.1% in all the bars and elevated copper contents (also around 0.1%) in Bars 76 and 78. Values for all other elements analyzed were below 0.1% (Table 4.7). The only notable heterogeneities in the metal compositions within and between bars is that of the carbon distribution, which was not quantified by EPMA, but assessed based on the metallographic examination, as well as in the presence of SIs (see below).

Metallography

The macrostructure and microstructures of the samples were studied to further understand their composition and how the iron bars were manufactured. The macrostructure of the samples can be seen in (Plates 20–22, Figs 4.15–4.19), accompanied by illustrated transects of the microstructure and optical micrographs (reflected light mode). All the samples studied represent complete or partial transverse cross-sections of the bars, except for Bars 18 and 46 where samples were obtained at the point where the bar bends (folds) longitudinally (perpendicular to the bar cross-section). The metallographic study of the Gresham bars revealed a great deal of heterogeneity within and between bars, which is clear in the sample sections and illustrated transects. A summary of the macro- and microstructures of the samples studied can be seen in Table 4.8. The carbon distribution and SI types identified in samples is provided in Table 4.9.

The sample sections, after etching, show clear contrasts in colour and brightness, as well as observed features (inclusions, weld lines and corrosion). In the study of the Roman iron bars from Saintes-Maries-de-la-Mer, it was noted that darker areas correspond to increased carbon content, whereas areas appearing bright generally represent lower carbon contents (Pagès et al., 2011, 1236). Although a similar observation was made during the study of the Gresham iron bars, the contrasts in brightness and colour (grey tones) observed macroscopically also coincided with changes in the microstructure such as grain size and enrichment bands. Therefore, the features observed macroscopically, including brightness and contrast, are representative of more than the differential carbon distribution. When discussing particular microstructures and features, supporting optical micrograph examples will be referred to by iron bar numbers, which are found in Figs 4.15–4.19 (Plates 20–22).

Microstructure

The majority of the bars exhibit a microstructure that is predominantly equi-axed ferrite (see micrographs from iron bars 22 and 40). This observation is consistent with raw bloomery iron. The equi-axed grains indicate that

GROUP	BAR	MACROSTUCTURE			MICROSTRUCTURE							
		layered	weld	slag	eq.	def.	hypo.	hyper.	wid.	regions	band.	enrich.
A	40	■			●		■			■		
A	42			■	●			■			■	
A	43	■			●			■	■	■	■	
A	44	■	■			●	■	■		■	■	■
A	45			■	■	■	●	■	■	■		
A	47	■			●	■	■	■		■		
A	78 S1		■	■	■		●	■		■	■	
A	78 S2				●	■						
B	18	■	■		●		■				■	■
B	76		■		●		■	■			■	
frag.	17	■	■		●		■	■	■	■		
frag.	22				●	■		■	■	■		
frag.	30			■	■	●		■	■	■		
frag.	38	■	■		●		■	■		■		
frag.	39				-	-	-	-	-	-	-	-
frag.	46	■	■		●	■	■	■			■	
frag.	48	■		■	●	■	■	■	■	■	■	

Table 4.8 Summary table of macro- and microstructure of iron bars. Key to abbreviations: macrostructure: layered = layered structure; weld = weld lines; slag = slag rich; microstructure: eq = equi-axed ferrite; def = deformed ferrite; hypo = hypo-eutectoid structures; hyper = hyper-eutectoid structures; wid = widmanstätten structures; regions = localised microstructural regions; band = microstructural lamellae banding; enrich = surface enrichment bands. Larger circles (●) indicate predominant microstructure

GROUP	BAR	CARBON DISTRIBUTION					SLAG INCLUSIONS	
		dist.	bands	regions	welds	SI	sing.	mult.
A	40	het.		■			●	
A	42	het.	■			■	●	■
A	43	het.	■			■	●	
A	44	het.	■			■	●	
A	45	het.		■				●
A	47	het.			■			●
A	78 S1	het.	■			■	●	■
A	78 S2	none					●	■
B	18	min.	■				●	■
B	76	het.	■					
frag.	17	het.	■	■	■		●	■
frag.	22	het.		■			●	
frag.	30	het.				■	●	
frag.	38	het.	■				●	
frag.	39	-	-	-	-	-		-
frag.	46	het.	■				●	■
frag.	48	het.	■	■			●	■

Table 4.9 Summary table of carbon distribution and types of slag inclusions present in iron bars. Key to abbreviations: dist = carbon distribution; het = heterogeneous; min = minimal; bands = well-defined hypo-/hyper-eutectoid lamellae bands; regions = diffuse/amorphous hypo-/hyper-eutectoid regions; welds = carbon distribution associated with weld interface; SI = carbon distribution associated with slag inclusions; Slag inclusions: sing = single phased; mult = multi-phased. Larger circles (●) indicate predominant SI type

the bars were annealed long enough for partial or complete grain recrystallization. The ferrite microstructure in Bars 30 and 44, however, shows flattened oval grains (no strain lines) that indicate plastic deformation after final heating. It is certainly possible that this resulted from the sampling process for these two bars. Bar 48 and one sample from Bar 78 are composed of equi-axed ferrite and interstitial pearlite consistent with a hypo-eutectoid microstructure.

Widmanstätten structures, remnant in most cases, can be observed in small areas of many of the specimens. They are most prevalent in Bar 30. This indicates that the bars were heated to high temperatures in the austenite temperature range (*c.* 850°C) followed by air cooling. The predominance of equi-axed ferrite indicates crystal regrowth as a result of annealing. This would also explain why the widmanstätten structures observed are localized to small areas or are remnant, where regrowth was incomplete.

Grain size varies a great deal within samples (see micrographs from iron bars 18, 38, 44, 47 and 78). Any annealing that was undertaken was not long enough for uniform crystal regrowth, hence the observed remnant structures. The heterogeneity is clearly defined by the alternating layers of pure (fine and coarse) ferrite and hyper-eutectoid structures, many of which are also delineated by elongated slag inclusions indicative of welds. Whilst these layered structures can be attributed to the folding and the welding of the iron, in most cases they are likely to represent the varied microstructure primary to the original bloom, subsequently elongated (to form bands) as the bar was drawn out. The banded structures are neither repeated nor regular throughout any of the sections, as would be expected if they were the result of secondary smithing. The fine grained ferrite may indicate a high degree of cold working prior to crystal regrowth. Considerably larger ferrite grains observed in some localized areas, particularly on surfaces, reveal a low level of deformation prior to annealing. Overall, the variation in grain size within samples is consistent with the heterogeneity observed in the carbon distributions and welds, which appear to be primary to the iron stock (see below).

Slag inclusions, weld lines and voids

The smithing of iron aims to consolidate the metal, shape it and expel slag inclusions. During the consolidation process of iron blooms, slag is extruded and iron surfaces welded together to form a solid metallic mass. Compared to well-consolidated high-purity iron, the Gresham bars can generally be described as slag-rich. This is particularly visible in Bars 17 and 42, where a 'vein-like' structure is apparent from the network of interconnected amorphous slag stringers, characteristic of unrefined bloomery iron.

Distinctive weld lines were observed in all of the samples examined, with the exception of Bars 22, 45 and 76. These can be observed macroscopically (see illustrations from iron bars 17, 18, 38, 44, 46, 47 and 48) and microscopically (see micrographs from iron bars 17 and 44). The welds were not always successful, in that iron oxide layers in conjunction with corrosionand slag can demonstrate cold shuts (see micrograph from iron bar 43). The welds generally appear intrinsic to the iron stock, formed during bloom consolidation. This is supported by the large slag stringers, often interconnected, that are associated with surrounding diffuse carbon regions. Many of the sample sections show voids present in the macrostructure, which are also visible microscopically (see micrographs from iron bars 30, 40, 42, 45 and 48). These voids are primary to the iron stock, showing that the iron had not been fully consolidated. This is consistent with the working of raw iron blooms that have only partly been refined.

Carbon distribution

None of the carbon distributions observed can be described as homogeneous. All pearlitic structures observed were localized or diffuse within certain areas of each sample (see micrographs from iron bars 17, 42, 45, 46 and 78). High-carbon regions and localized hyper-eutectoid regions were identified in many of the bars, which in some cases were associated with weld seams (Bars 17 and 47) or SIs (Bars 30, 42, 43, 44 and 78). Ten of the samples showed either discrete or obvious banded carbon structures, resulting from the elongation of carbon-rich zones with the work-thinning of the iron bars. Where the layering is clearly pronounced with sharp boundaries, it can be attested to folding and welding of the iron prior to being incorporated into the stock as mentioned earlier. The segregation of the carbon in some of the banded structures may also be related to internal variation of the metal composition. This is supported by the appearance of enrichment bands, visible as white lines in the microstructure (see micrographs from iron bars 18 and 44). The heterogeneous distribution of carbon, with some prevalence near SIs and intrinsic weld seams, indicates that the carburization is primary from the smelt and not the result of any intentional ironworking processes. This is further supported by the coarse appearance of the observed pearlite, where cementite/ferrite lamellae are often irresolvable, indicative of slow-cooling and a feature of bloomery iron. The microstructure of the Gresham iron bars, therefore, indicates them to be ferritic iron, with the exception of Bars 45 and 78, which could arguably be identified as mild steel.

Two bars (42 and 76), however, do have a hypo-eutectoid layer on their outer surface (see micrograph from iron bars 46 and 76), diffusing inwardly to the ferrite substrate. It might be tempting to interpret this structure as resulting from a form of 'case-hardening', where the iron bars have intentionally been carburized. However, for true case-hardening, one would expect a deeper penetration of the carbon into the iron bars, rather than the thin superficial layer observed here. These iron bars, therefore, might have had the appearance of steel (if tested at their surfaces only), masking their real constitution. Whether this is intentional or not remains speculative, especially when compared to the rest of the other bars.

Shaping process: power-hammering

The square-section bars (Type A) are not uniform in their thickness. They generally taper down from their mid-section (thickest) towards their tips (thinnest). The bar thicknesses can vary from 40 x 40 mm at its middle, down to 20 x 20 mm (the accuracy of the bar measurements may be slightly affected by corrosion). Indentations are visible on many of the square-section bars, resembling a series of concave-convex waves that repeat every 40–60 mm (Plate 23, Fig. 4.20). The undulations are consistent with a wind- or water-powered hammer – a tool well attested historically and archaeologically in later medieval European forges associated with both bloomeries and blast furnaces (Awty, 2006, 136; 2008; Hayman, 2005, 18; Smith, 1997, 37; van Laun, 1979, 57–9).

Although commonly associated with blast furnace technology, automated hammers were already being used to process iron blooms from the direct process. In Europe, the use of trip hammers in forges appears around the start of the 12th century AD (Needham, 1986, 379). An invaluable insight into how the Gresham iron bars were fashioned can be gleamed from near contemporaneous artistic depictions and historical sources. Book IX of Agricola's *De re metallica* provides several references to forging iron bars as well as two woodcuttings of tail hammers used to forge iron bars (Plate 23, Fig. 4.20). What is particularly worthy of note is the mention of square bars, which we can assume to mean square-sectioned, based on the woodcutting depictions printed in the text:

> These pieces [of iron bloom], after they have been re-heated in the blacksmith's forge and again placed on the anvil, are shaped by the smith into square bars or into ploughshares or tyres, but mainly into bars (Hoover and Hoover, 1950, 423).

In 1555 in Rome Olaus Magnus published a book including woodprints of aspects of culture and history of the Nordic countries, with special reference to Sweden. One of the woodcuts shows three water-powered hammers beside a set of bellows operating a furnace (Fig 4.20); this has been interpreted as an Osmund furnace (Needham, 1986, 395). A very similar arrangement can be seen in a painting by Jan Brueghel the Elder (1568–1625) in the Museum of Fine Arts of Lyons; entitled 'Fire' (1611), it depicts the scene of 'Venus in Vulcan's forge' and shows an iron bar being drawn using two water-powered hammers (Fig 4.20). The historic Frohnauer hammer mill still survives today in south-east Germany and is dated to 1660, but was in use nearly half a century earlier. It has three water-powered hammers set up in the same fashion as depicted in Magnus's woodcut print.

Most of the Gresham iron bars appear to have been fashioned by a similar type of hammer. The indentations themselves are not only negative impressions of the hammer in use, but also a product of the hammer action and weight. The head must have been flattish with slightly rounded edges, no doubt from wear.

The bars all share a similar feature, which is the thinning that occurs at the folds themselves. Although visible in many of the bars themselves, the feature is best seen by X-ray. X-radiographs of Bar 48 and parts of Bar 40 are annotated in Fig. 4.21 (Plate 23), where hammer indents are marked out, as well as the thinning of the bars located at the bends. This means that the bars were worked thinner at specific points prior to folding, where bending was intended. No doubt this would ease the process of folding or 'packing' the iron bars down.

In contrast to the square-section bars, the flat bars classified as Type B bear no hammer marks. They are virtually smooth in comparison to Type A. The thickness of Type B bars is almost uniform throughout, except for the ends, where they become marginally thinner. It is unlikely that such a finish was achieved by hand. More likely is another power-assisted process to fashion the final bar, but the lack of hammer marks make it possible that these bars were rolled.

Both 'squares' and 'flats' appear to have been widespread in the international iron market. A manuscript dating to around the 1660s and thought to be written by Swedish ironmaster Abraham Cronström (1640–1696) (Pipping, 1982) notes that such bars were made in Sweden, but also in Spain (Basque Country and Catalonia), the Netherlands (Liège), France (Lorraine), Germany (Harz region and Westphalia), Wales and England, and although the discussion is focused on blast furnaces, references are made to the direct method in Catalan and German bloomeries (Björkenstam, 1982, 36–9).

Standardization and bar iron quality

The information presented so far highlights the relatively low standardization of the Gresham bars, especially when compared to later examples of bar iron including those mentioned in historical texts (see below). No correlations can be established between different features of the iron bars, such as length and weight. Equally, there is no correspondence between the finishing method (hammering or rolling) and the carbon content of the iron, or even the smelting system it derives from (manganese rich or poor). Also variable is the grain size and the distribution of carbon and slag inclusions within and between bars. The only consistent features in the Gresham bars are the thickness and width within each of the 'square' or 'flat' groups, as well as the size of the 'packets', the folded lengths, which are around 2 m or, perhaps more appropriately, 6 ft. All of these aspects will be recaptured in subsequent discussion, together with the notion of the 'quality' of the metal.

Even though we cannot judge iron qualities by today's standards, relevant parameters can be gleamed from older references showing that producers and consumers could discern 'sound iron'. Rather than typifying iron into 'better' or 'worse' quality, it may be preferable to discuss iron in terms of different types 'suited to different uses' (Hildebrand, 1992, 64). Barraclough (1984, 37) cites an early 18th-century textbook to distinguish types of bar iron available, the five categories being: 1) tough and soft; 2) tough and hard; 3) cold short; 4) hot short; 5) brittle, being both hot short and cold short. This categorization takes us beyond mere aesthetic appearance to the material properties of the iron itself.

Another two criteria for judging the quality of iron are derived from the work of Swedish chemist and mineralogist Sven Rinman (1720–1792) and relate to the nature of the slag inclusions and consistency (Hildebrand, 1992, 63). Good bar iron should be slag-free with the same quality throughout. Although this statement pertains to iron produced by the indirect process, it is worth noting how German bar iron was regarded as being softer and of inferior quality compared to Swedish Walloon iron, which was harder and 'sounder'. This was reflected in the price. The German smiths, however, were highly regarded for their ability to produce iron according to exact dimensional specifications (Hildebrand, 1992, 63–5).

The scant archaeological evidence for post-medieval trade in bar iron supports the existence of variable qualities. Near the isle of Texel at a site at Aanloop Molengat, a Dutch shipwreck (probably having sunk around 1635) was excavated, carrying a cargo of metals including iron bars largely consisting of squares (35 mm) and flats (60 mm x 15–20 mm) (Maarleveld and Overmeer, 2013). One of the iron bars (produced by bloomery) was sampled, exhibiting a cross-section containing no more than 1% surface area in slag inclusions and a steely encasing with a ferritic core (Maarleveld and Overmeer, 2013, 118). This would suggest a high-purity mild steel that is practically slag-free – a stark contrast to the Gresham bars which are comparatively slag-rich with a heterogeneous microstructure.

Iron provenance hypotheses based on slag inclusion analysis

As described above, the analysis of the iron bar slag inclusion chemistry revealed two main groups, possibly related to production provenance. It could not, however, indicate what those provenances might be. Provenance identifications, regardless of technique or material require two distinct datasets. The first is a characterization of material from potential production sources (for example, sites, site clusters or regions with known locations in space and, preferably, time). These constitute training sets. The second dataset consists of material with unknown origins. The attribute patterns of the unknowns can then be compared to those recognized in the training sets to identify potential matches and mismatches.

The chemical attributes of bloomery iron objects, particularly those of their slag inclusions (SIs), provide the most precise means of identifying their production origins. The isotopic and trace element chemistry of the metal is linked directly to the ore from which it is smelted (Degryse *et al.*, 2007; Degryse *et al.*, 2009; Devos, *et al.*, 2000), as is the trace and major elemental chemistry of SIs (Coustures *et al.*, 2003; Desaulty, *et al.*, 2009; Dillmann and L'Héritier, 2007; Hedges and Salter, 1979; Leroy *et al.*, 2012). Unlike the metal, however, smelting slag chemistry (from which most SIs are derived) is also influenced by the composition of furnace construction materials and fuel ash (Crew, 2000). The study of experimental slag and SIs by Blakelock *et al.* (2009) showed that these additional influences of the smelting recipe could also be important for identifying production sources.

The first step in a provenance analysis is to identify and demonstrate differences between potential sources represented by the chemical composition of smelting slag or ores. Methods include univariate (Buchwald and Wivel, 1998), bivariate (Blakelock *et al.*, 2009) and multivariate (Birch and Martinón-Torres, forthcoming; Charlton, *et al.*, 2012; Hedges and Salter, 1979; Leroy, *et al.*, 2012) analysis of element and oxide distributions. The fields created by these data have known patterns of dispersal in spacetime and become origin or provenance hypotheses. The second step in a provenance analysis is to project new chemical data from metal or SIs into the space created through characterization of the potential sources. Matching distributions of source and object data indicate that a provenance hypothesis is not rejected while misaligned distributions of source and object data lead to the rejection of a provenance hypothesis.

While such approaches are common for the provenance study of non-ferrous metals, as illustrated above for the lead and tin, similar methods for the provenance of iron are much less developed. The provenance of the iron bars recovered from the Gresham Ship was investigated with the multivariate strategy proposed by Charlton *et al.* (2012). This strategy uses Linear Discriminant Analysis (LDA) to reveal the major structural differences between *a priori* defined groups (sites, site clusters, or regions) characterized by the major and minor element chemistry. Similarly characterized SIs from iron objects can then be projected into the multivariate space created by the analysis of the smelting slag groups to assess provenance hypotheses.

The morphology of the bars, coupled with the approximate age and location of the wreck, indicate that they were produced somewhere in Britain, Scandinavia, or the Rhineland. However, trade from other regions like Spain were not discounted. In ideal circumstances, large smelting slag samples would be collected from these areas and their compositions measured via instruments that had been cross-calibrated. Such endeavours, however, rest outside the constraints of the present project. Assuming that the geochemical fingerprints of regional slag assemblages will maintain broad differences between groups despite variation in instrument and time, then it is possible to create training sets from published slag chemistries. Provenance fields, so constructed, can be used to weight source hypotheses, but not to falsify them.

A search for suitable smelting slag compositional analyses from published sources resulted in the production of twenty training sets covering broad areas of southern Britain, the Netherlands, Scandinavia and the Basque Provinces of Spain. Slag analyses from an iron production site at Wolkenberg, Southern Brandenberg, Germany, were also included in the comparison. A reference summary for each training set is given in Table 4.10.

Six oxides (MgO, Al_2O_3, SiO_2, K_2O, CaO and MnO) were chosen for the LDA of the training sets based on the few missing values reported in the published smelting slag datasets and no missing values amongst the SI datasets. Following procedures detailed in Charlton *et al.* (2012), all data were transformed to the negative logarithms of their subcompositional ratios and a LDA conducted on the training sets. Provenance fields were generated by a Kernel Density Estimation using the Sheather-Jones direct plug-in method for bandwidth estimation. The fields are defined by the minimum contour lines containing 100 % of the training set scores within a given perspective of the LDA space. Finally, the SI data were projected into the LDA space and their position compared to the provenance fields. Inclusion of a single SI within a provenance field was deemed sufficient to retain the field as a likely origin hypothesis. All analyses were conducted with the software package R version 2.15.2 (R Core Team, 2012) using functions from the base and MASS (Venables and Ripley, 2002) packages.

Inspection of the projected SI values onto the first three LD axes led to the rejection of eleven training sets as likely provenance hypotheses. A second iteration of the analysis led to the rejection of a further four training sets. A third iteration added negative weight to two more competing provenance hypotheses. The remaining three included West Denmark, the Midlands Jurassic region in England and the Veluwe region in the Netherlands. A fourth analysis was conducted to clarify their hypothesis potential. A graph of the bar SI data projected into the LD space (Plate 24, Fig. 4.22) reveals a poor correspondence of the bars to any of the three provenance fields. The low Mn SIs plotted nearest to the Veluwe field, but with few plotting within the field itself. Only one SI plotted in the Midlands Jurassic field and none in the West Denmark field. These results indicate that the true provenance of the bars remains undetermined, but that greatest weight should be given to the Veluwe provenance hypothesis and, by extension, to a Rhineland hypothesis for bars with low Mn SIs. The bars with high Mn SIs have no clear provenance association.

Training sets	n=	Rejection stage	Source
1. West Denmark	20	4	Buchwald (2005)
2. Oppland/Jamtland (Norway And Sweden)	7	1	Buchwald (2005)
3. Fyn, Denmark	4	1	Buchwald (2005)
4. Sjaelland, Denmark	5	1	Buchwald (2005)
5. Sweden	10	2	Buchwald (2005)
6. Beitostolen, Norway	4	1	Buchwald (2005)
7. Sysendalen and Hardangervidda, Norway	15	1	Rosenqvist (1983)
8. Forest Of Dean And Bristol Mendip Carboniferous, England	50	1	Paynter (2006)
9. Kent Tertiary Sands, England	20	2	Paynter (2006)
10. Surrey And Hampshire Tertiary Sands, England	10	2	Paynter (2006)
11. Sussex Wealden, England	26	1	Paynter (2006)
12. Midlands Jurassic, England	12		Paynter (2006)
13. Llwyn Du, Wales	51	1	Charlton et al. (2010)
14. Vecht, Netherlands	24	1	Joosten (2004)
15. Veluwe, Netherlands	17		Joosten (2004)
16. Montferland, Netherlands	6	1	Joosten (2004)
17. Wolkenberg, Germany	32	1	Spazier (2003)
18. Alava, Spain	9	2	Gil et al. (2008) and I. Yusta (pers. comm.)
19. Bizkaia	17	3	Gil et al. (2008) and I. Yusta (pers. comm.)
20. Gipuzkoa	9	3	Gil et al. (2008) and I. Yusta (pers. comm.)

Table 4.10. List of slag data training sets employed in the LDA, identified by location with accompanying references

This finding, however, must be tempered with several caveats. The first two were mentioned above and involve problems of instrumental variability and the use of smelting slag of dissimilar age to those of the bars. None of the training sets used here are the same age as the bars. Indeed, they are between 300 and 1500 years older. Using non-contemporaneous training sets denies the possibility of chemical change as a consequence of technological evolution or shifts in resource utilization. A third issue concerns the use of small samples to define training sets, most of which contain less than twenty slag analyses. Such small sample sizes are unlikely to provide an accurate representation of chemical variability. A fourth point relates to the structure of the training sets, none of which were systematically constructed with provenance analysis as a stated goal. Instead, they represent groupings based on political geography or proximity to particular ore bodies. The only exceptions to this pattern are training sets derived from single sites and those constructed from Buchwald's catalogue of Scandinavian smelting slag (2005). Fifth and finally, the number of variables is used in this study is very small. There is little doubt that the incorporation of more oxides or trace elements would lead to better discrimination between training sets.

With so many caveats constraining the results, it is easy to dismiss the conclusions of the provenance analysis as meaningless. The position taken here, however, is that although the results are not conclusive they do indicate locations where future provenance research might be most fruitfully applied.

Traffic and trade

Voyage iron?

To summarise, we have established that the iron was made by the bloomery process and subjected to relatively limited consolidation and refining by hot forging prior to being shaped into bars. Most of these were shaped by welding several blooms together with the help of a power hammer, although two of them may have been rolled. The bars are generally standardized in their cross-sections (whether flats or squares) and in their lengths as packed at around 2 m, but they are quite variable in their internal quality – especially in the carbon distribution and the conspicuous presence of imperfect weld lines and slag stringers. The chemical analysis of the slag inclusions has established that the bars derive from two smelting systems. That said, the close morphological similarity between the bars from the two chemical groups suggests that they could have been manufactured in the same area. Although unproven, it is not impossible that all the iron examined could derive from the same smelter. The provenance study, although hampered by biases in the reference data, points to the Rhineland as the most likely source for this metal – at least, for the Mn-poor group.

There are numerous historical references to the production and trade of iron – often in bar form – across the world since the late Middle Ages. Some of the more famous types clearly differ from the Gresham Ship material in composition, quality or method of production and therefore in the markets they were destined for. Öregrund bar iron, produced in Uppland, Sweden, for example, and highly reputed for its quality and toughness (Evans and Rydén, 2007, 161; Hildebrand, 1992, 63), and Osmund iron, another Swedish product known to England from the late 13th century (Calissendorff, 1979, 173; Magnusson, 2011b; Nyquist, 2011) were, unlike the Gresham bars, produced by the indirect method. Hoop iron and slit rods, which were sometimes rolled into *rutheneisen* – rod iron – (Björkenstam, 1982, 48–49; Hildebrand, 1992, 67–68), are much too small to be compared to the likes of the Gresham bars. 'Tilted steel', famously produced in Styria (Austria) and imitated in Sweden and Germany, sometimes took the form of bars produced, perhaps similarly to the Gresham bars, by the hot forging and folding of a bundle of crude bars (Björkenstam, 1982, 41) – but the product is said to have 'the evenness required of a good steel' (Björkenstam, 1982, 41), far superior to the microstructures observed in our metallographic study.

There is, however, one type of bar iron that may be of more direct relevance to our study. The so-called 'voyage iron' or 'Guinea iron' was produced primarily for export to West Africa, where it was often traded for slaves. The most famous producers of iron specifically aimed at the African market were the Swedish and the term is often taken to imply a Swedish origin (Evans and Rydèn, 2007). However, while according to documentary sources the production and trade of Swedish voyage iron appears to have taken off only from the mid- to late 17th century, the African metal trade had begun in the previous century with the Portuguese. Trading voyages and raids by Dutch, French and English vessels, which had been only sporadic during the earlier part of that century, intensified from the 1580s. Bar iron is mentioned as one of the prime goods brought to Africa by these European merchants (Alpern, 1995, 20; Hawthorne, 1999; Brooks, 2003, 69). The Bristol Port Books for 1600/1 show a few references to 'Guinea iron', providing documentary proof of the early export of iron from Britain to West Africa (Jones, 2009). By the mid-17th century, we have evidence of English traders based in Sierra Leone and extending to the north (Brooks, 2003, 103–4). It is also at this time that the English involvement in the transatlantic slave trade appears to heighten (Evans and Rydén, 2007, 163), but we should not forget that the process had already begun by at least 1562, with the infamous voyage of Captain John Hawkins, regarded as one of the first, if not the first, slave traders.

It is therefore not impossible that the Gresham Ship bar iron represents an early example of a potential trading voyage destined for the West African market – but are there any documentary data to support the possibility of iron being imported into England from the Low

Countries for that trade? Peter King (2005) has produced the most comprehensive account of the evolution of iron production and imports into England since the late medieval period. Based on his data, Fig. 4.23 (Plate 24) shows a timeline of English bar imports. During the lifetime of the Gresham ship (*terminus post quem* 1574), Spanish imports had already declined considerably prior to the intermittent Anglo-Spanish war (1585–1604) and the Eighty Years' War (1568–1648) in the Low Countries. Incidentally, imports from the Baltic appear to peak around this time, although these imports are comparable in total tonnage to that continuing from Spain. Importantly, there is a significant period of growth in English iron production from the 1540s to the 1580s, both using the direct and indirect process; for nearly 50 years up until 1620, England became almost self-sufficient (up to 90%) in iron (King, 2005, 16, 20). This is significant, as it certainly limits the likelihood of the exported Gresham bars being English products. Indeed, it has been claimed that 'the majority of the bar iron that left England was 'foreign merchandise exported', that is re-exports (King, 2004, 249).

King provides relatively little quantitative information related to the import of iron from the Rhineland prior to the later 17th century. This may be a reflection of the relatively minor weight of Dutch iron in English imports during the period. However, that such a trade did take place in the 16th century is confirmed by the Port Books for 1567/8, since, for example, three vessels that brought iron from Antwerp to London are recorded in that year: one of the consignments is specifically listed as 25 cwt (1270 kg) of iron from Liège (Dietz, 1972, nos 18, 45, 466).

Alpern (1995, 12) states that the bar iron traded in West Africa was made in 'England, Sweden, Spain, France, Belgium, Germany, Denmark, Norway, Ireland and even Siberia.' Cox and Dannehl (2007, *s,v.* voyage iron) specifically state that English traders re-exported voyage iron from the Rhineland and the Liège district – although specific dates are not given. According to Evans and Rydén (2007, 164), voyage iron does appear to have been traded by Dutch merchants at least as early as the beginning of the 17th century. Thus such historical information lends some support both to the concept of a Dutch origin for the re-exported bar iron and also the possibility of their use as an early form of 'voyage iron'.

The next question concerns whether the quality and morphology of the Gresham bar iron was consistent with such a trade. On this front, the main problem is that most of the specific information related to voyage iron comes from Sweden and post-dates the wreck assemblage. In the 1660s, Cronström (1640–1696) makes a very pertinent statement in his records about trading poor quality iron in connection no doubt with the slave trade:

> Similarly, one chooses the poorest iron one can find for sending to Guinea. But it must be forged into a form that is in demand. This also applies to iron that is shipped to the Caribbean region and other islands, as the East and West Indies. They never ask for good iron there (Cronström, 1982, 194).

Cronström also compares Swedish iron to that of the Low Countries, stating that, 'Swedish iron ... when hot, is harder than Liège iron which seems easier to work when hot' (1982, 166). If the iron from the Low Countries is easier to work, it implies it is soft (i.e. low carbon or pure iron), as is the case with the Gresham bars.

A description of voyage iron by the Swedish chemist and mineralogist Sven Rinman in his *Dictionary of Mining* (*Bergverks Lexicon*, 1788–9, *s.v. resa järn*) states that voyage iron bars are, 'bent double two or four times, so that in foreign places [they] can be carried on a donkey'. The Gresham bars, some bent double and others doubled twice, would appear to conform to this description. However, the size and weight of the Gresham bars appear large when compared to those in (later) historical descriptions. For example, Cronström states that:

> Voyage iron must be 11 ft (3.4 m) long and must amount to 18–20 bars per 5 cwt (226.8 kg), or 76 to 80 in a cask of 20 cwt (907.2 kg), quite smooth and well forged. A bar of good flat iron [should be] 1⅞ inch (48 mm) wide, $^{7}/_{16}$ inch (11 mm) thick, 15 ft (4.6 m) long ... (Cronström, 1982, 189, 197).

The weight stipulated by Cronström of between 72–80 bars to the ton equates to 11.3–12.6 kg each. This accords with a record of bar iron weighing 70 bars to the ton, as supplied to the Dutch West India Company in the mid-17th century (Evans and Rydén, 2007, 165). The weight of voyage iron appears to progressively decrease during the latter part of the 17th century (75–80 bars per ton) and early part of the 18th century (92 bars per ton), as pointed out by Evans and Rydèn (2007, 165). In comparison, almost 15 bars to the ton would be obtained using the much heavier Gresham examples. Their weight corresponds better with the *waag*, a German weight unit of 120 lb (54.4 kg) (Pipping, 1982, 254).

The complete bars among the Gresham assemblage do range in length from around 12 to 19 ft (3.7 m to 5.8 m) which would also seem to exceed the minimum length that qualifies for voyage iron. Cronström's description of Swedish voyage iron also provides slimmer characteristics than the Gresham bars. However, Hildebrand indicates that squares could range from ⅞″ to 2″ square (22–51 mm), sometimes even being as small as ¾″ (19 mm) (1992, 66). This could imply that the square dimensions of voyage iron were somewhat flexible and dependant on a client's wishes or a buyer's preference. In fact, the dimensions of iron bars are rarely specified (Björkenstam, 1982, 42) and, in some cases, they were made to specification – what would become known as 'iron made to pattern' (Hildebrand, 1992, 67). In sum, the dimensions and weight of bar iron only appear to be of critical importance by the end of the 17th century (Evans and Rydén, 2007, 165).

The documentary record therefore suggests that voyage bar iron became progressively smaller and more standardized over time but, as noted above, none of these sources refer specifically to the period and possible provenance relevant to the Gresham iron. It would seem plausible that, when this branch of trade in bar iron was beginning, other bar typologies, sizes and weights could have found their way to Africa. The Gresham Ship may thus have been bound for Africa with a consignment of voyage iron, but alternative destinations for the mixed cargo she carried also merit serious consideration – especially after the rest of the cargo is brought back into the picture.

Tin, lead and red lead: alternative hypotheses

There is comparatively little evidence for lead and tin being carried to West Africa in this period. Alpern (1995) mentions both metals in his catalogue of goods traded for slaves in 'Kwaland' (the shoreline from Liberia to Nigeria, so called by Alpern after the Kwa language of the local peoples), but acknowledges that there are only 'a handful of references' to tin and the lead he records is in the form of bars and probably from Portugal. Furthermore, he is not precise with the dates. The Oranjemund shipwreck was carrying copper, lead and tin ingots, as well as well as a substantial amount of elephant tusks and other commodities, when it sank off the Namibian coast in the 1530s (Chirikure *et al.*, 2012b). This could demonstrate that non-ferrous metals were brought to Africa and that European merchants stopped at various places on route to engage in the trade of various commodities. However, it is thought that the final trade destination for this vessel was in the Indian Ocean and is thus unlikely to provide a direct comparison for the Gresham Ship. About a century later, the Aanloop Molengat Wreck sank while carrying lead, tin, mercury, iron bars and cloth. Maarleveld and Overmeer (2013, 142) suggest that this vessel may have been bound to trade 'through the straits of Gibraltar', possibly with stopovers en route where convenient. Interestingly, however, the size and shape of the iron bars are consistent with those of 17th-century 'voyage iron'. In 1682 an eye witness report stated that, in Senegambia, Europeans exchanged slaves for 'cotton baft, copper, tin, iron, spirits and a few glass trinkets' (Barry, 1998, 49).

If the Gresham ship was involved in slave trade, it is also possible that the lead and tin were not destined for Africa but for the New World, perhaps together with the African slaves obtained in exchange for the iron. American colonies were reliant on non-ferrous metals from Europe (see, for example, Hudgins *et al.*, 2009). Lead was necessary in America for construction and warfare, but also for the extraction of silver from American ores (Thibodeau *et al.*, 2007). Thus, while it is plausible that all three metals on the Gresham Ship were destined for the West African trade, a Caribbean or American destination cannot be ruled out, at least for part of the cargo – the evidence is by no means conclusive.

It is at this point that we need to consider a further commodity that may have been part of the cargo, even if there is no archaeological evidence for it – red lead. The 19th-century report of the cargo partly salvaged from a shipwreck that seems highly likely to have been the Gresham Ship, includes reference to 'red lead in cast iron casks, covered with wood' (British Archaeological Association, 1847, 361). Red lead, Pb_3O_4, also known as 'minium' is a form of lead oxide typically obtained by calcination of PbO or litharge. Derivatives of lead have been used for medical purposes since Antiquity, but the main use of minium, before its industrial use as an anti-corrosion agent, was as a pigment. Red lead was not only used as a cheaper alternative to cinnabar for artistic painting, but also, together with white lead, on a larger scale for the painting of buildings.

It is obviously impossible for us to prove where the red lead in the Gresham ship might have come from, but the main producers of English red lead for both domestic use and for export were in the Peak District (Kiernan, 1989). Interestingly, this is the same source as the lead ingots from the Gresham Ship. Of even greater significance is the fact that Derbyshire red lead production really only begins on a major scale in 1590 (Kiernan, 1989). Thus the presence of English red lead in the cargo would imply a date for the shipwreck towards the turn of the 16th to 17th century, as opposed to the 1570s, when the ship was built. Such a date would also be consistent with the dates inferred from the iron trade and with the relative longevity suggested by the extensive repairs documented in the ship's hull.

Although voyage iron is strongly related to the Africa trade, as described above, no references have as yet been found that mention red lead being exported there. Alternative destinations for the cargo should therefore be considered in addition to Africa and the New World, including our neighbours in northern Europe. The Bristol Port Books contain a reference to a ship by the name of *Grasshopper* (the creature that was, incidentally, the emblem of the Gresham family), which was importing iron into Bristol in 1575/6 (Jones, 2009): the same ship also carried lead to France in 1576. As Bristol is located near the heart of lead and tin production in England (Cornwall), it is not surprising that English ships were exporting such non-ferrous metals from that port. Indeed, the Bristol Port Books for 1600/1 show substantial exports of lead to the continent too, including France and also Italy, but there is little mention of tin (Jones, 2009).

In the particular case of the Gresham Ship, it is unlikely that iron was imported from the continent to be re-exported back there. However, the above paragraph highlights the fact that we should be open to other possibilities in addition to Africa or America. This raises the possibility of the Levant Trade for the Gresham Ship, a voyage that of necessity ran past Gibraltar into the Mediterranean. This avenue is discussed in more detail in Chapter 5. Epstein (1908, 232–8) recorded the impositions charged on 87 categories of goods imported by the English Levant Company into Turkey. The notes are based on lists compiled since 1608 but almost certainly reflect earlier trading patterns. Interestingly, these categories include iron, tin, lead and also red lead. Even though these records do not specify the shape of the iron traded, bar iron would potentially be useful for construction reinforcements and other applications in any areas. Although the best studies of architectural bar iron available come from France, where long bars of iron, often of mediocre purity are mentioned (L'Heritier *et al.*, 2010), it is conceivable that similar bars were produced in England for international trade through the Levant Company. Arguing against this possibility, however, is Willan's (1955) study, indicating that the balance of English trade with the Levant was unfavourable and showing evidence for the export of English tin, but virtually none for iron.

Conclusion

The typological, microstructural, chemical and isotopic study of the metals from the cargo, coupled with historical research, has allowed us to suggest with various degrees of confidence the possible technological and geographical origins of the lead, tin and iron that sank with the Gresham Ship. We have proposed that, while the lead and tin are most likely English, the iron bars may constitute import-to-export commodities sourced from the Rhineland. Considering the variety of metals aboard and their seemingly varied sources, there is little doubt that the vessel was sailing outbound when it floundered in the estuary. Furthermore, the study of the cargo offers several pointers to strengthen the proposition that the Gresham Ship was relatively long-lived, since both the production of red lead and the export of bar iron possibly for the West African trade are more consistent with a late 16th- or even early 17th-century date.

It is impossible at this stage to establish precisely where the cargo was destined for. While voyage iron remains a strong possibility, connecting the ship with West Africa and perhaps with the New World, further possibilities cannot be ruled out completely – notably including Levantine trade through the Mediterranean Sea. The best that can be said of the Gresham cargo (or that proportion we know of) is that it was almost certainly destined not for a short-haul Channel crossing, but for a relatively distant shore, as has been argued for the Aanloop Molengat wreck. As such it represents the widening horizons of English merchantmen in this period; it was not bound for Antwerp, but for Constantinople, Africa or even America.

Figure 4.1 Bottom and top views of lead ingot from the Gresham Ship after mechanical cleaning of marine encrustations

Figure 4.2 Detail of one of the stamps identified on surface of lead ingots

Figure 4.3 Metallographic structure of lead ingot under plane polarised light, showing clusters of lead sulphide (grey), one of them surrounded by lead oxide (black). Field of view is ~400 μm

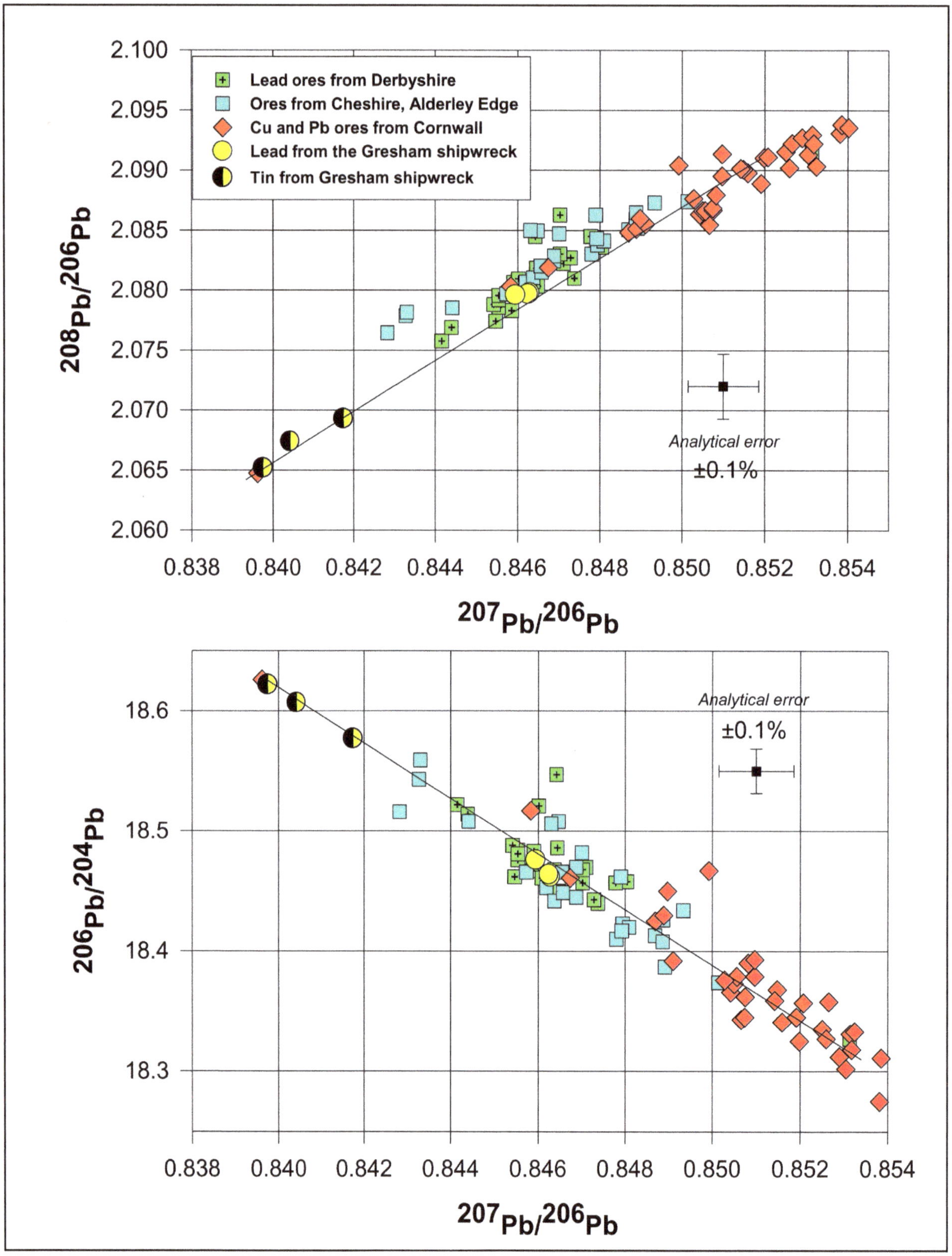

Figure 4.4 *Two 2-dimensional mirror plots of lead isotope ratio data for two lead ingots (the two samples from ingot L2 overlap completely) and three tin ingots from the Gresham ship, compared to ores from Derbyshire, Cheshire and Cornwall. The line joining the points for samples of copper and lead ores from Cornwall represents the 'mixing line' of lead from copper and lead ores possibly present as contamination in the tin ingots. The analytical error of ±0.1% is the overall analytical error of lead isotope measurements by Thermal Ionisation Mass Spectrometry (TIMS) measured by repeat analyses of lead isotope standard NBS 981 over a long period of time (Gale and Stos-Gale 2000). Since the lead isotope comparative database of ore samples is mainly based on the TIMS data this seems an appropriate analytical error for the comparative plots*

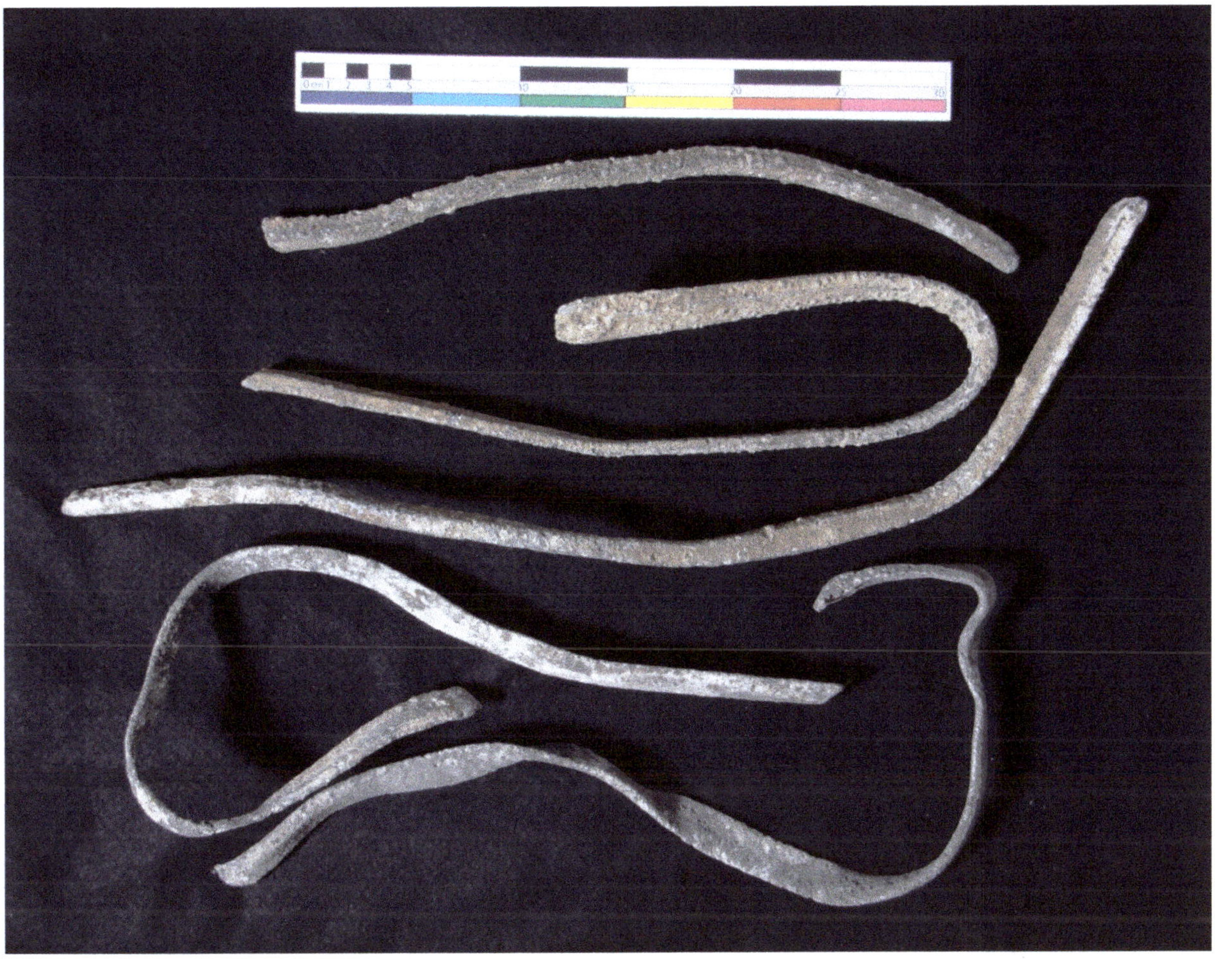

Figure 4.5 Tin ingots recovered from the Gresham Ship

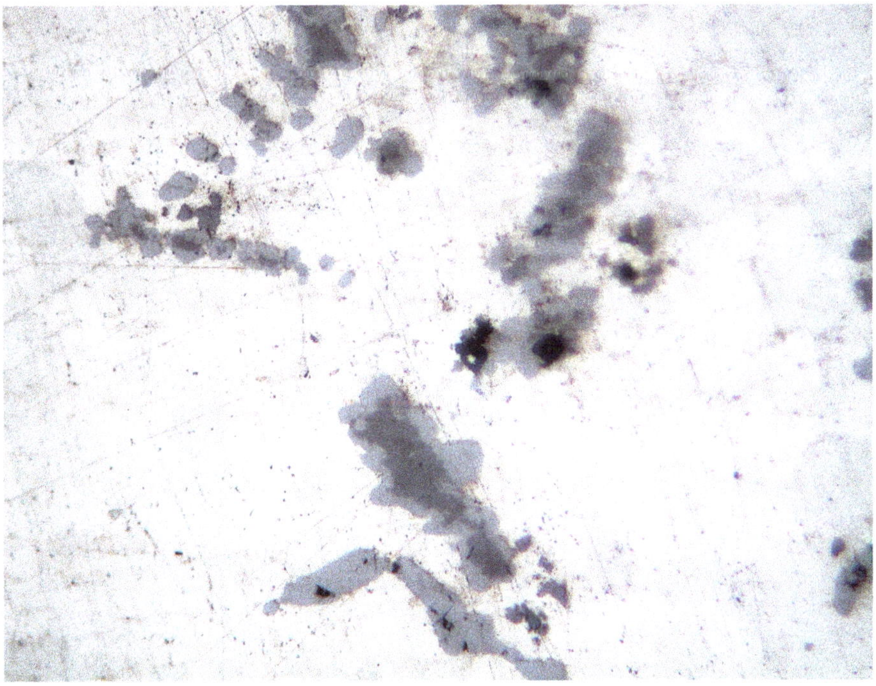

Figure 4.6 Metallographic structure of tin ingot under plane polarized light, showing crystals of FeSn (light grey), some of them with a darker core of Fe_3Sn_2. *Field of view is ~200 µm*

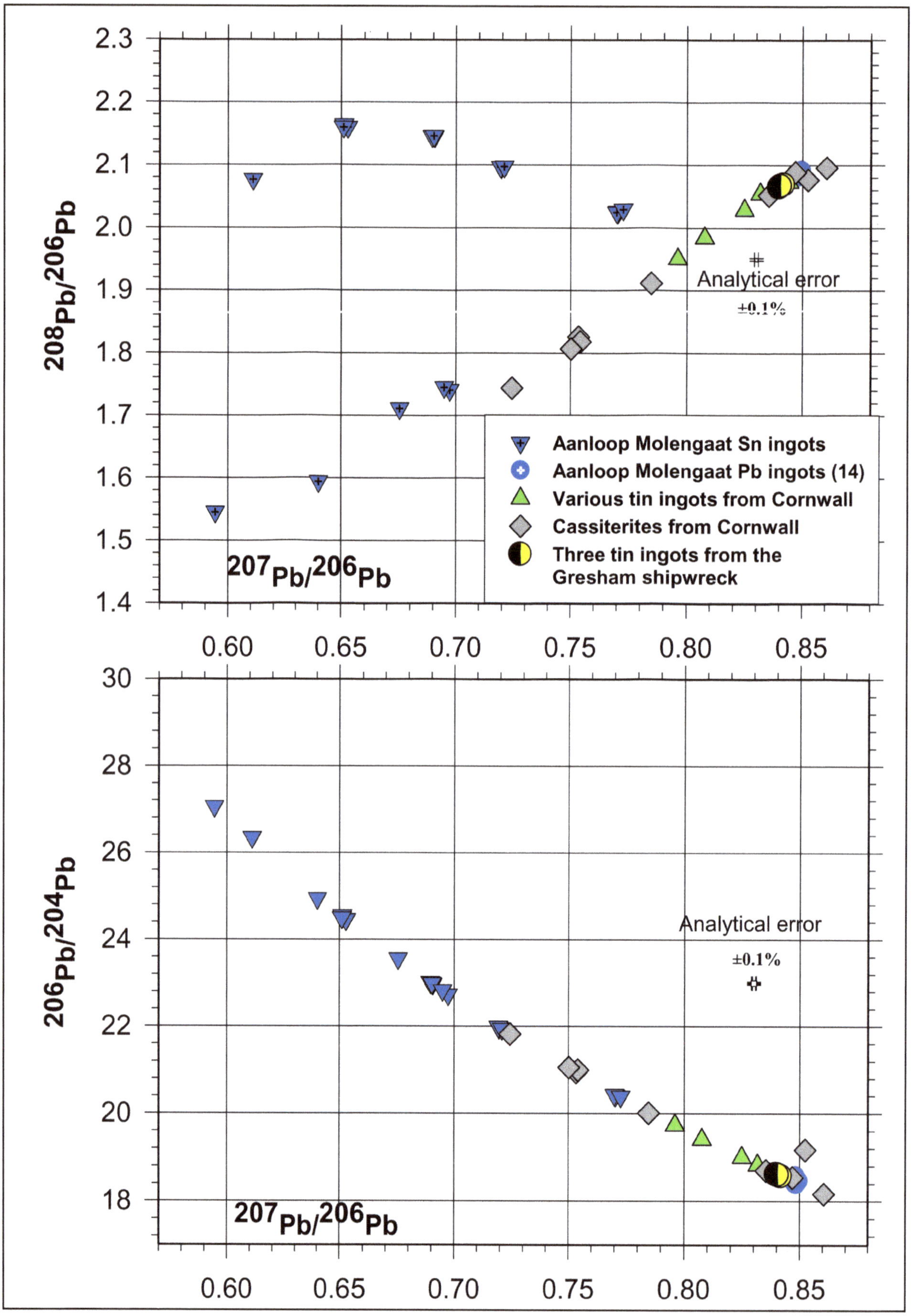

Figure 4.7 *Two 2-dimensional mirror plots of lead isotope ratio data for tin ingots from the Gresham Ship, compared to cassiterites from Cornwall, as well as the lead and tin ingots from the Aanloop Molengaat wreck. Tin ingots from Anloop Molengaat bear heraldic impressed seals which the excavators have used to deduce that the ingots were made from tin ores from the Erzgebirge in Saxony. This comparison shows that the tin carried on the Gresham ship is of different origin*

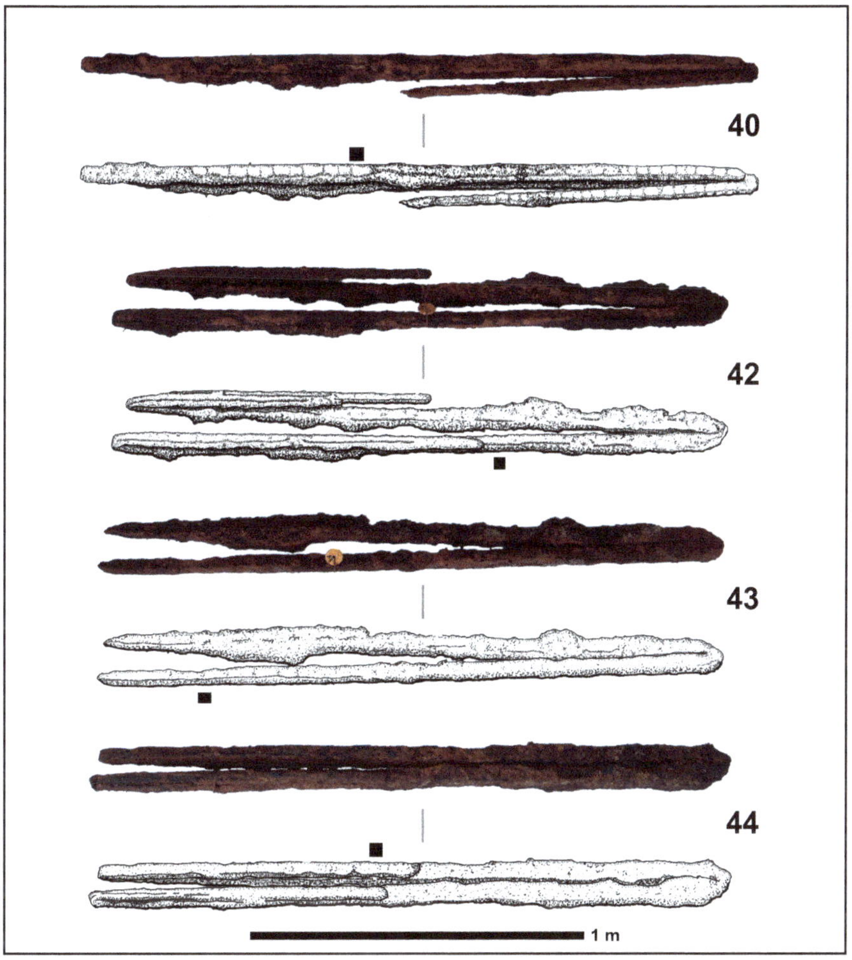

Figure 4.8 Photographs and illustrations of 'squares' (folded iron bars 40, 42, 43 and 44, square cross-section)

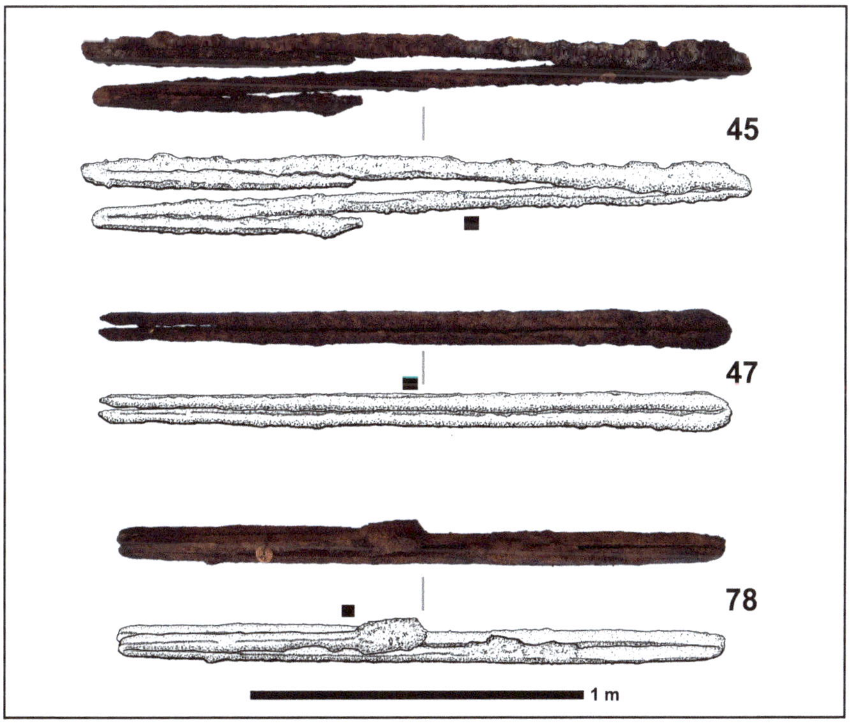

Figure 4.9 Photographs and illustrations of 'squares' (folded iron bars 45, 47 and 78, square cross-section)

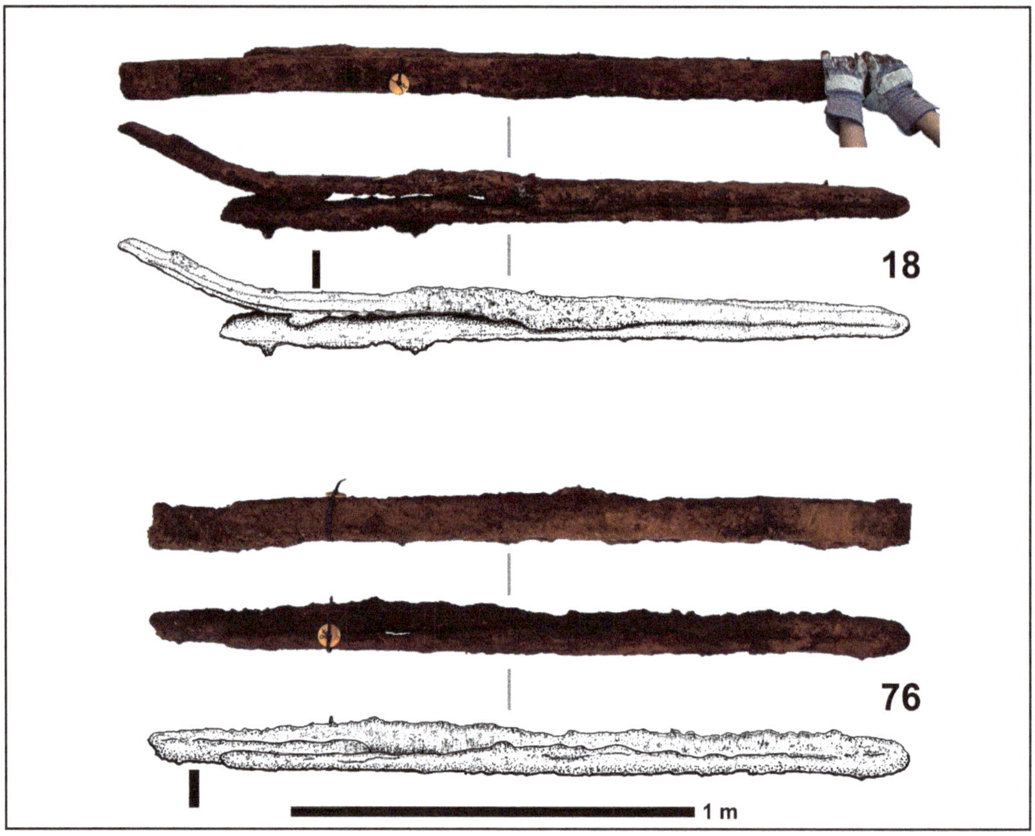

Figure 4.10 *Photographs and illustrations of 'flats' (iron bars 18 and 76, rectangular cross-section)*

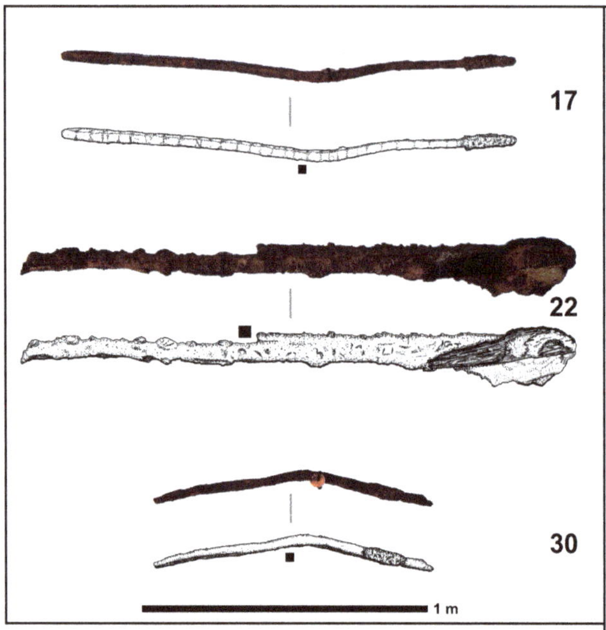

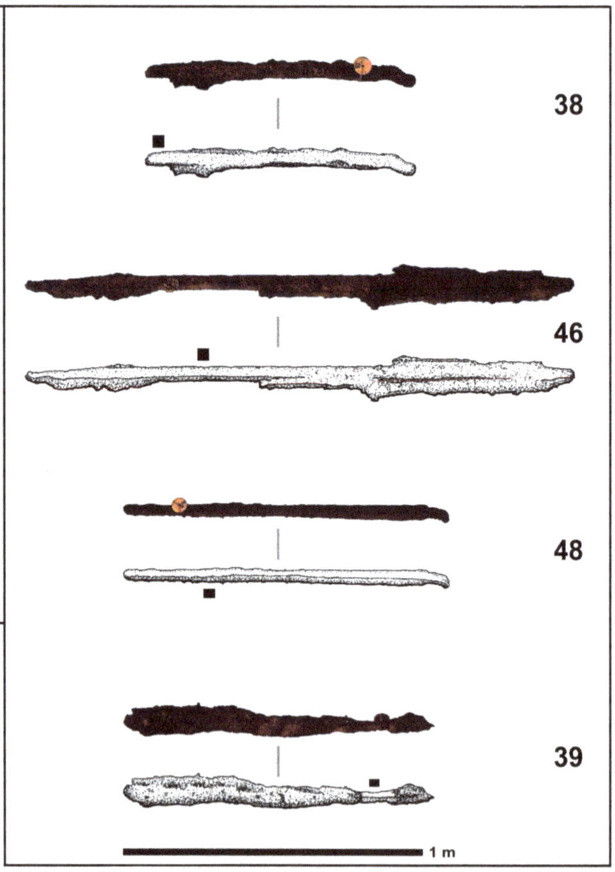

Figures 4.11 (above) and 4.12 (right) Photographs and illustrations of non-diagnostic fragments, iron bars 17, 22 and 30, and 38, 39, 46 and 48 (square cross-section)

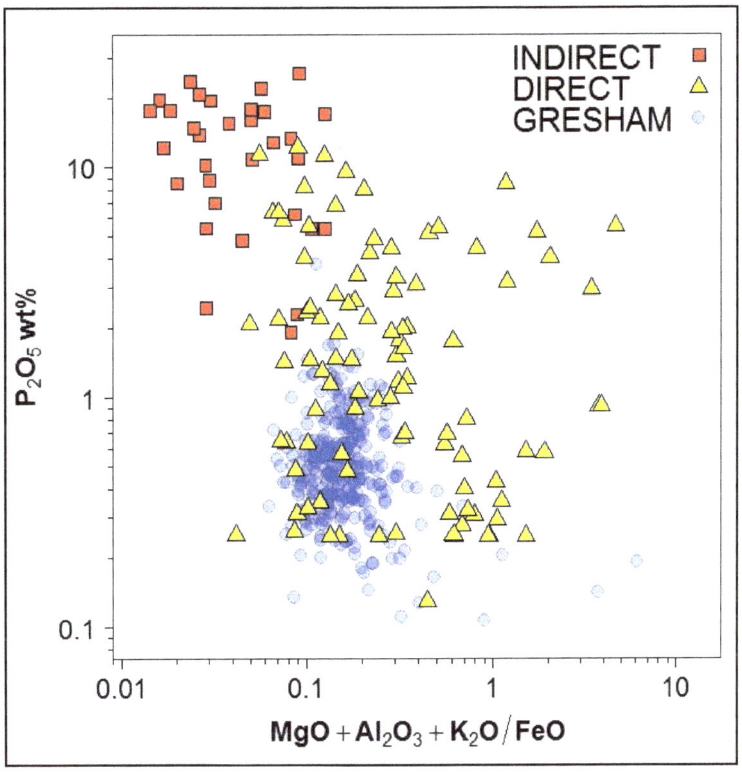

Figure 4.13 Bivariate scatterplot distinguishing direct and indirect processes based on the ratio of phosphorous pentoxide to weighted contents of several other compounds in the SIs. Reference data from experiments provided by Philippe Dillmann. Iron bar SI EPMA-EDS data are plotted as transparent markers to illuminate point density. Axes are log-scaled to improve visibility of data point clusters

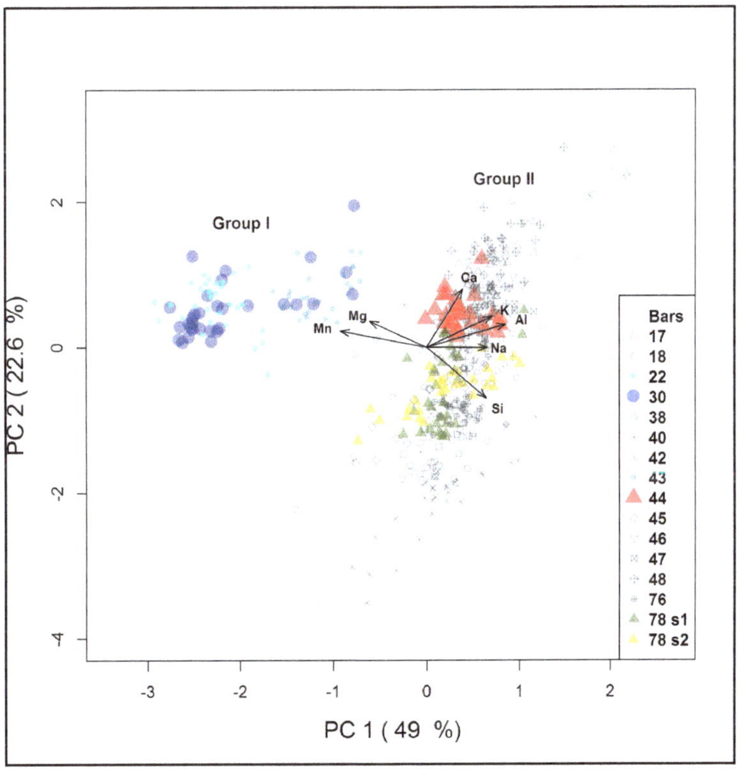

Figure 4.14 Plot of Gresham iron bar SI compositions projected on to principal component axes 1 and 2. Distribution of SI scores plotted by bar reveal the presence of two main chemical groups and a tendency for SIs in each bar to form identifiable clusters within the broader group

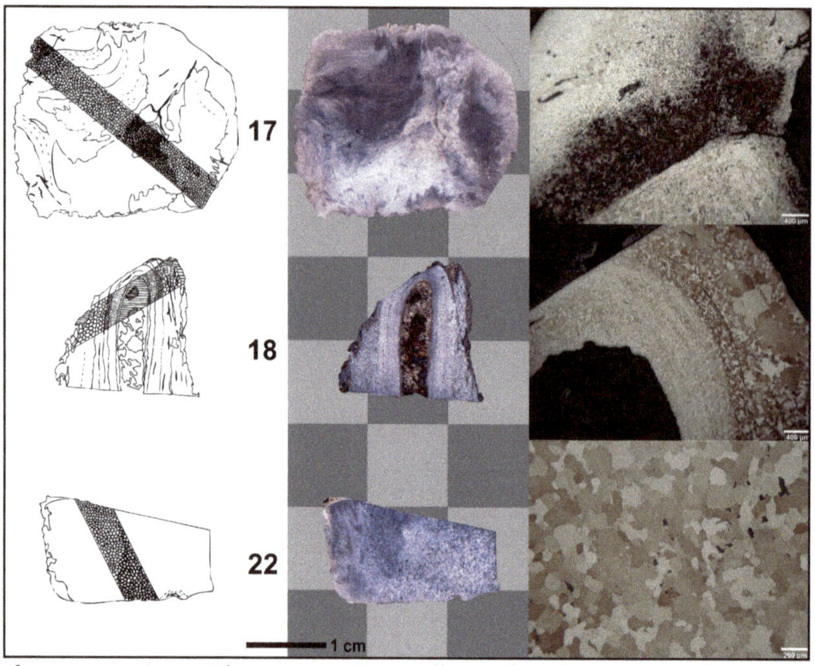

Figure 4.15 Sample cross-sections with macroscopic metallographic illustration and micrographs of Bars 17, 18 and 22.

Bar 17 macro: visible banded structures, large primary weld seam and high carbon region, micro: localized carbon concentration (eutectoid, 0.8% C) on one side of a weld-seam, bordered by equi-axed ferrite;

Bar 18 macro: layered structure and visible interconnected slag inclusions, micro: large recrystallized ferrite grains (right) bordering layers of small grained ferrite containing enrichment bands (white lines) and slag inclusions (directional quality);

Bar 22 macro: single body of iron with diffuse boundary from higher carbon region and some irregular banding, micro: mostly recrystallized equi-axed ferrite with some irregular grains, with region containing slag inclusions

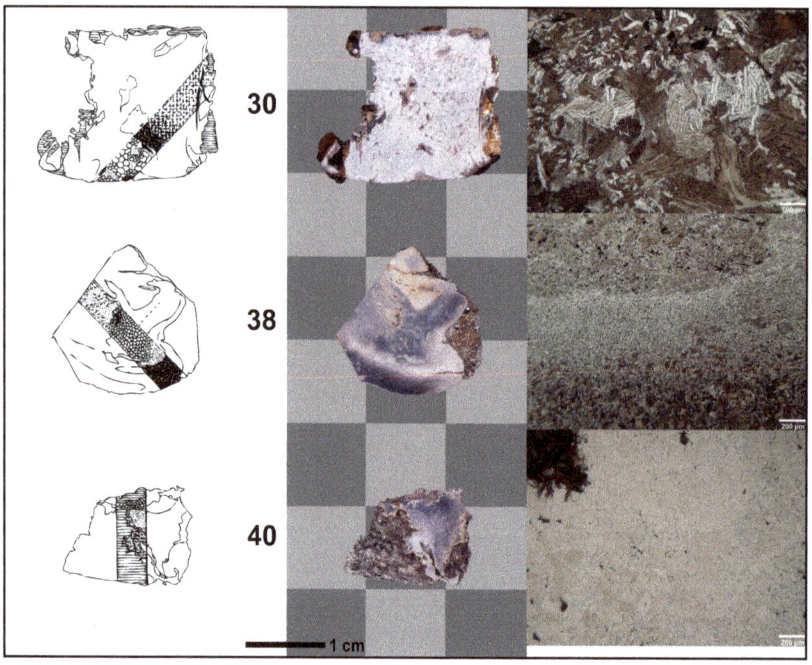

Figure 4.16 Sample cross-sections with macroscopic metallographic illustration and micrographs of Bars 30, 38 and 40.

Bar 30 macro: non-uniform patchy appearance with obvious irregular voids, micro: widmanstätten ferrite with interstitial pearlite (<0.2 % C);

Bar 38 macro: two clear weld-lines (one coinciding with carbon rich band) and multiple irregular shaped inclusions (corrosion and voids), micro: diffuse carbon gradient (eutectoid to hyper-eutectoid, 0.8% to 0.2% C) and change in ferrite grain size, broadly corresponding to two different layers;

Bar 40 macro: one possible primary weld (discontinuous) with localized diffuse higher carbon region, micro: equi-axed ferrite grains with some porosity (void with embedded corrosion)

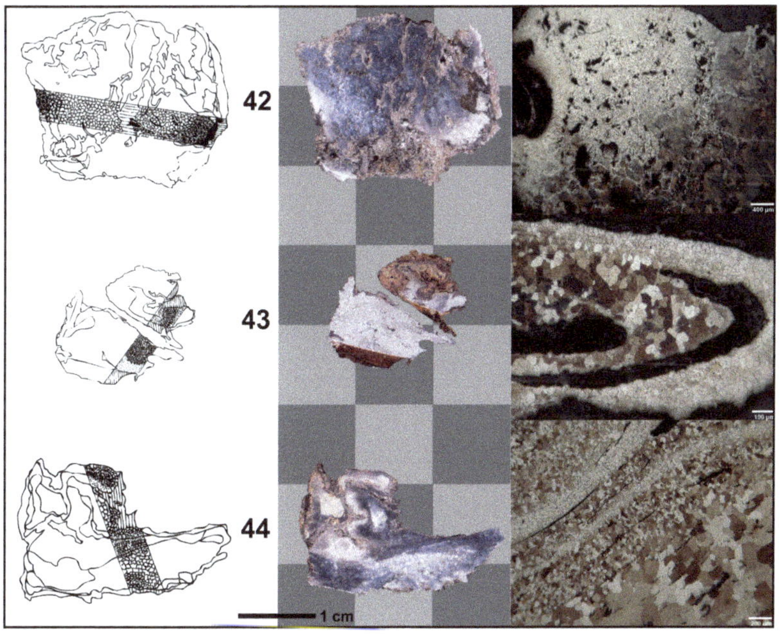

Figure 4.17 Sample cross-sections with macroscopic metallographic illustration and micrographs of Bars 42, 43 and 44.

Bar 42 macro: large network of interconnected inclusions (slag rich), with obvious voids with corrosion, and two higher carbon concentrations, micro: ferrite region bounded by hyper-eutectoid region (0.8–1% C), identified by pearlite and some cementite (visible as precipitates on grain boundaries and widmanstätten side plates);

Bar 43 macro: irregular distribution of microstructures with an example of banding, micro: bent iron mass constructed of two bands of different grain sized ferrite separated by cold shuts (incomplete welds) visible as iron oxides;

Bar 44 macro: distinct primary weld lines with multiple localized concentrations of carbon, micro: enrichment bands (white lines) respecting elongated slag inclusions indicative of welds, corresponding with changes in grain size (equi-axed ferrite) and carbon segregation (interstitial pearlite, <0.2% C)

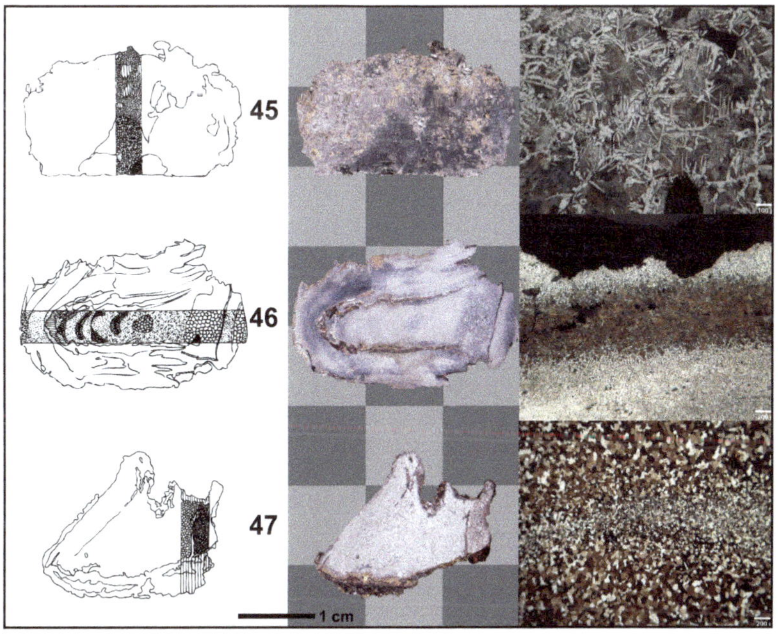

Figure 4.18 Sample cross-sections with macroscopic metallographic illustration and micrographs of Bars 45, 46 and 47.

Bar 45 macro: densely filled with inclusions (voids, corrosion, slag) and characterised by diffuse boundary between two regions (one carbon rich), micro: several voids in coarse pearlite with ferrite precipitating on grain boundaries and as widmanstätten side plates (0.6% C);

Bar 46 macro: two weld lines, one being the axis of the contoured layered structure with visible banding; micro: ferrite with band of pearlite (eutectoid, 0.8% C) near the sample surface;

Bar 47 macro: two clear weld lines, one defining sharp boundary change in the microstructure; micro: small grained hyper-eutectoid band contained within equi-axed ferrite (0.2–0.3% C)

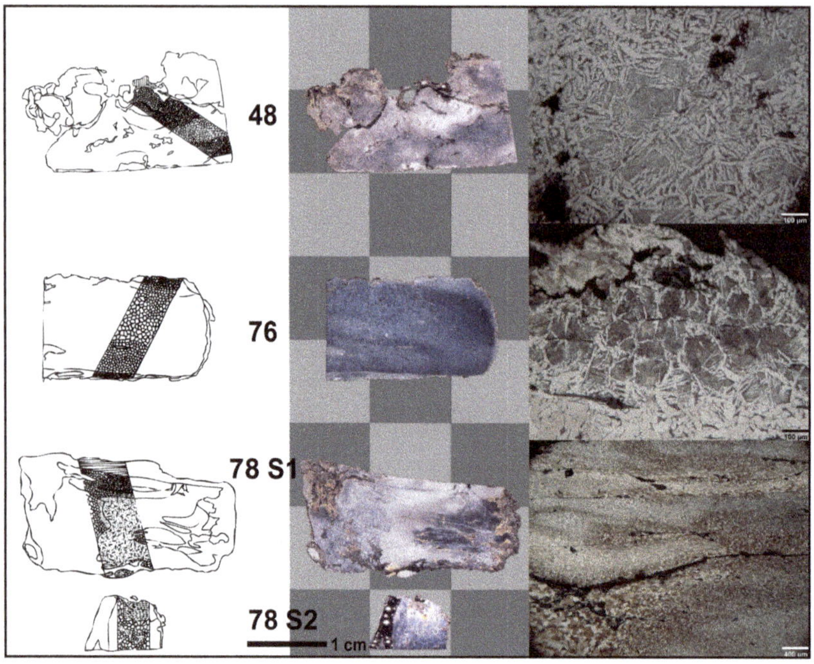

Figure 4.19 Sample cross-sections with macroscopic metallographic illustration and micrographs of Bars 48, 76 and 78.
Bar 48 macro: irregular concentrations of carbon, two primary welds and several voids, micro: irregular voids (bounded by corrosion) in coarse pearlite and widmanstätten ferrite (0.45% C) with diffuse boundaries into smaller grained ferrite with interstitial pearlite;
Bar 76 macro: single body of iron containing some inclusions (directional) with higher carbon layers located near the surfaces, micro: coarse pearlite and widmanstätten ferrite layer (0.6% C) bounded by ferrite;
Bar 78 macro: irregular distribution of carbon, interconnected networks of inclusions (slag, voids and corrosion) exhibited in both samples, micro (S1): sharp differences in the microstructure defined by interconnected network of slag inclusions where layers of ferrite have sharp boundaries with fine grained hyper-eutectoid bands (<0.2% C)

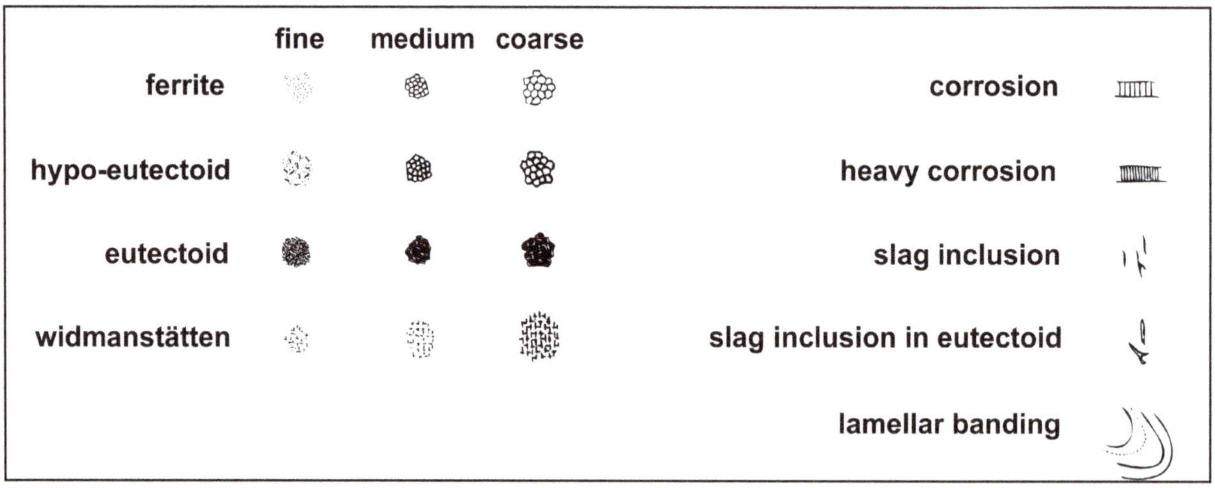

Legend for Figs 4.15 to 4.19

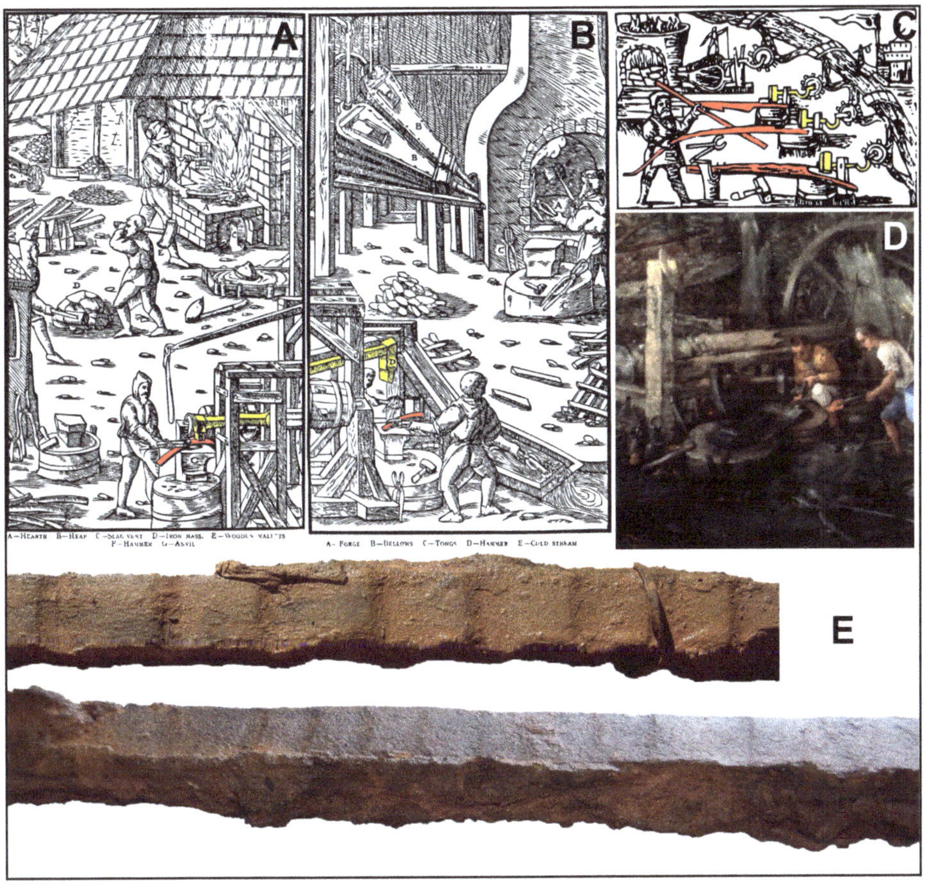

Figure 4.20 Late Medieval representations of tail hammers and the production of iron bars:
A) woodprint after Agricola (Hoover and Hoover, 1950, 422)
B) woodprint after Agricola (Hoover and Hoover, 1950, 425)
C) woodprint after Magnus (1555, 207)
D) after Jan Brueghel the Elder (excerpt from 'Fire', dated 1611, held at the Museum of Fine Arts of Lyon, image obtained from: http://commons.wikimedia.org/wiki/File:Fire-Brueghel_the_Elder-MBA_Lyon_A75-IMG_0409.jpg)
Hammers are highlighted in yellow and iron bars in red in the line drawings
E) close up images of hammer-mark indentations from Bar 40 (top) and Bar 44 (bottom)

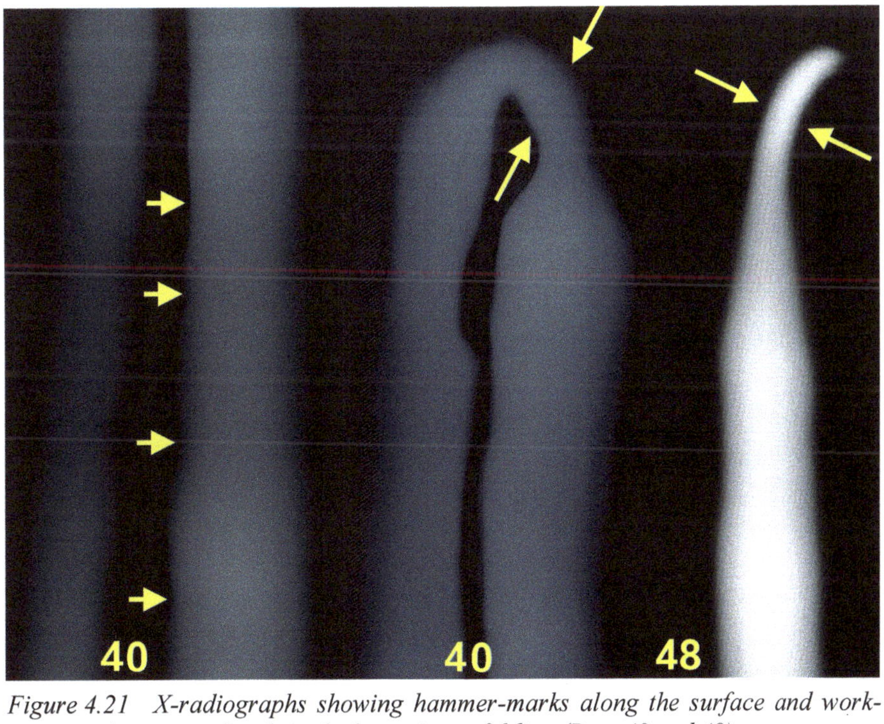

Figure 4.21 X-radiographs showing hammer-marks along the surface and work-thinning at bends in the bar prior to folding (Bars 40 and 48)

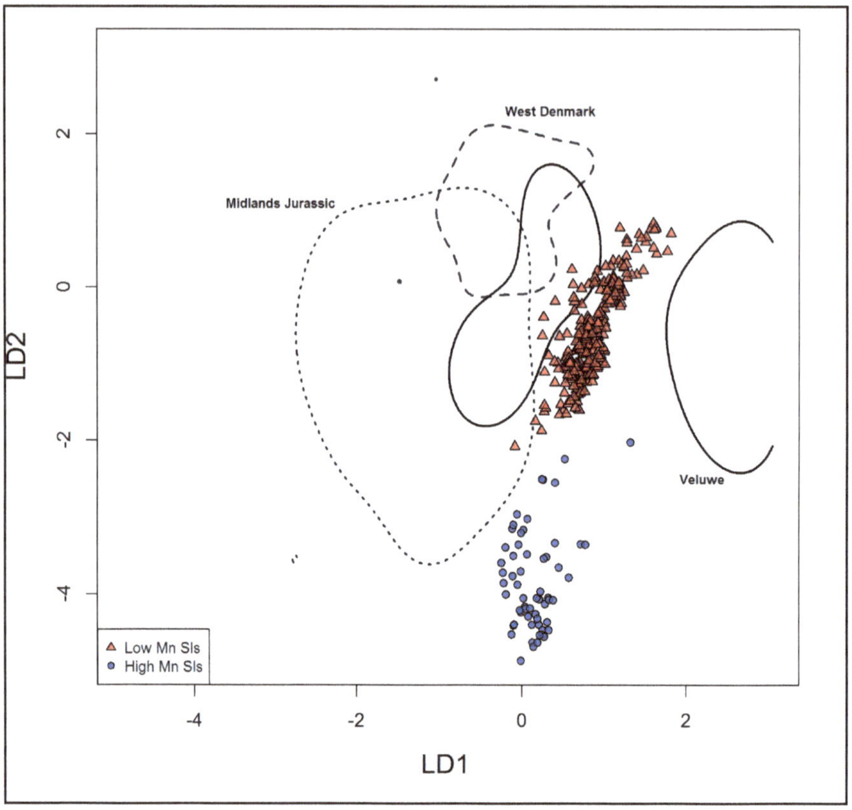

Figure 4.22 Final multi-group provenance evaluation of production sources from Britain, Denmark and the Netherlands using LDA. Provenance fields are defined by the minimum KDE contour level containing 100 % of training set scores. Data with missing or zero values were omitted from analysis

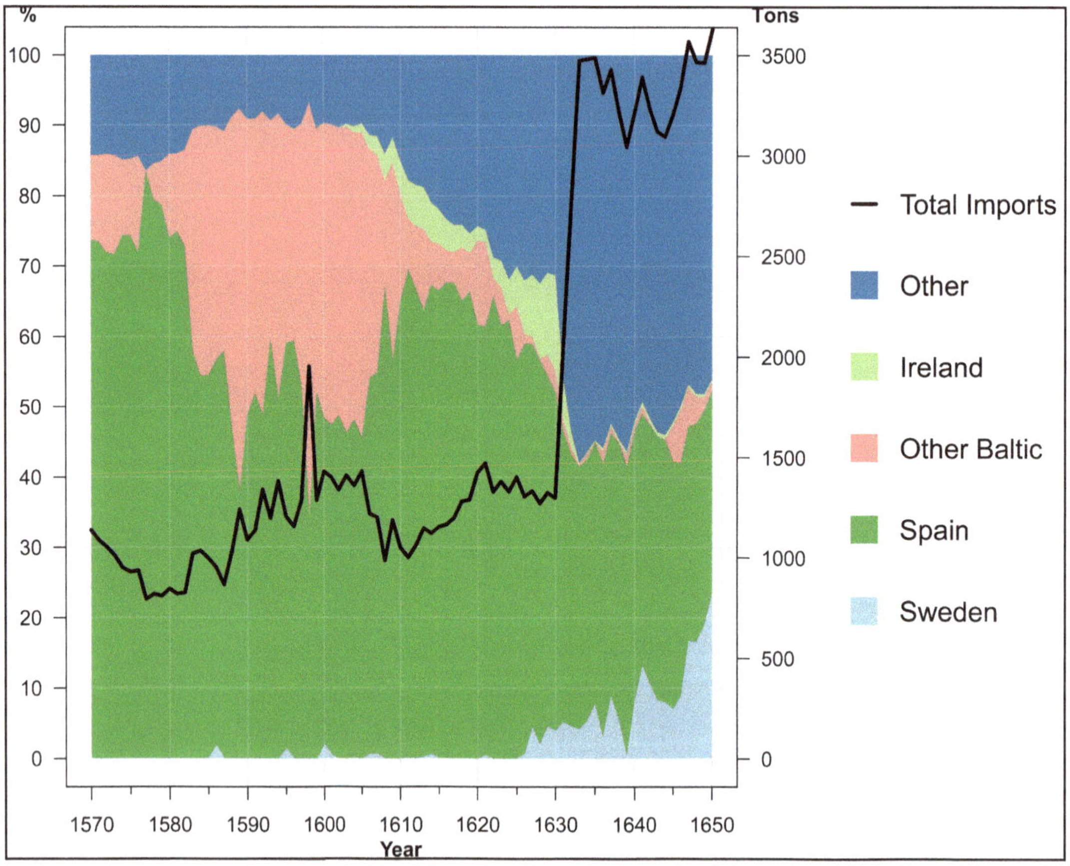

Figure 4.23 Historical timeline of iron imports in England AD 1570–1650, based on data compiled by Peter King (2004, Appendix 19). Black line represents total tonnage in absolute numbers, whereas colours show relative contribution of different origins

Chapter 5: Armed Elizabethan Merchantmen

by Gustav Milne

Trading and Raiding

The Gresham Ship is described as an 'armed merchantman', that is, a privately-owned trading vessel designed to carry cargoes, but also able to protect itself from attack in an uncertain world (see, for example, Bindoff, 1950, 247–76). Since it has not been possible to provide a secure identification for the Gresham Ship and therefore to compile a detailed history of its working life, this chapter summarizes some of the activities experienced by English merchant shipping and their crews in the late 16th century, to provide a contemporary maritime and mercantile context for our vessel. In this period Europe was riven by deep religious and political divisions, while piracy on the high seas was a commonplace, often feeding on those deep divides. The balance of naval power in the Mediterranean changed dramatically in 1571 with the defeat at Lepanto of the Ottoman Turks by a confederation of Christian fleets drawn from Italy and Spain. To the north the uneasy relationship between Roman Catholic and Protestant states manifested itself in diplomatic rows, guerrilla warfare or all-out invasion. In the 1560s, for example, Dutch merchantmen turned to aggressive privateering in response to the occupation of their lands by Spanish forces. For trading vessels, even in times of so-called peace, an attack could come from any quarter, the more so if the cargo carried was particularly valued. Consequently merchantmen had to be prepared to give as good as they got. Such ships would therefore have a complement of guns and the crew would include at least one professional gunner to supervise the maintenance and use of the ordnance.

This period also saw continuing developments in the design and construction of ships in the heart of the Mediterranean, on Europe's Atlantic seaboard and in the Low Countries, as shipwrights, merchants and monarchs responded to a world that was rapidly changing politically and geographically (Unger, 1980, 252–78). In England the warships being built for Queen Elizabeth were changing in the light of experience and with a view to different future. Matthew Baker and Peter Pett were the master shipbuilders at the heart of the innovations and the treasurer of the navy was John Hawkins, a highly experienced mariner and a man who 'knew how to equip a navy at a price England could afford' (Kelsey, 2003, 81). These documented developments included not just changes to the lines of a ship, but the tarring of hulls and the introduction of the chain pump (Kelsey, 2003, 160). It might be anticipated that some of these practices could also be appearing in contemporary merchant shipping, such as our Gresham Ship, a vessel built at the same time as such famous warships as the *Dreadnought*, the *Handmaid*, the *Revenge*, the *Triumph*, the *Victory* and the *White Bear*, for example (1573–8).

That said, English merchantmen continued to show pronounced design differences from a royal warship, 'race-built' solely for action (see, for example, Parker, 1996, 286). An early 17th-century treatise on building warships comments on the main differences, noting that 'merchants covet to have great floors in their ships for gaining of stowage, but thereby they spoil the ship's bearing for most of them grow tender-sided, and after they are built come to be furred' (Salisbury and Anderson, 1958, 16). William Borough, the Comptroller of Elizabeth's navy, identified three distinct vessel-types:

1. the most profitable trading vessel would be of 'the shortest, broadest and deepest order'. The length of the keel would be double the breadth amidships, while the depth of the hold would be half the breadth.
2. 'the mean and best proportion for shipping merchandise, likewise very serviceable for all purposes': The length of the keel two or two and a quarter that of the beam, with a depth of the hold $^{11}/_{24}$ that of the beam (i.e. just under half)
3. 'the largest order for ships for the wars made for the most advantage of sailing'. The length of the keel was three times the beam, with a depth $^{2}/_{5}$ of the beam (Oppenheim, 1896, 126)

In spite of such differences in size, design and purpose, merchantmen were nevertheless more than capable of fighting alongside the 'Queens Ships', should the occasion arise. Indeed in Elizabethan England merchant ships, their crew and many Thames watermen were regularly requisitioned for military purposes to augment the modest but bellicose 'Navy Royal' (Loades, 2009, 128–9). A contemporary view of this situation is recorded by John Stow, the London antiquarian and member of the Mercer's Company, writing on the City's maritime industries less than a decade after the defeat of the Spanish Armada: 'By the benefit of the river of Thames and great trade of merchandise, it is the chief maker of mariners and nurse of our navy: and ships, (as men know) be the wooden walls for defence of our realm' (Wheatley, 1956, 296).

The crown offered a financial inducement of five shillings a ton for the building of any privately-owned merchant ship over 100 tons, materially expanding those wooden walls for a modest capital outlay. But it is not clear how or if the bounty was actually paid; sometimes it seems as if the required sum was simply deducted from the merchant's subsequent tax bills. In

Ships over 100 tons built in London or east coast ports 1574–1579			
Year	Ship's name	Estimated tonnage	Associated shipyard or port
1574	*Christopher*	120	Ipswich
	Joseph	120	Cley
	Bark Bonner	130	Deptford
	Jesus	130	London
	Phoenix	130	Harwich
	Mary Fortune	140	Aldeburgh
	Salmon	140	Leigh
	Anne	150	Orwell
1575	*Goodgrace*	120	Ipswich
	Elizabeth	125	Leigh
	Henry Francis	140	Aldeburgh
	Ann Galant	160	Yarmouth
	Jonas	160	Ipswich
1576	*Judith*	110	London
	Margaret	160	London
1577	*Jesus*	140	Lynn
1578	no ships listed below 240 tons		
1579	*Jewel*	180	London

Table 5.1 Listing of newly built ships over 100 tons and therefore eligible to claim Royal Bounty (after Dietz, 1991, 5–20). All these vessels would be contemporary with the Gresham Ship; one might actually be the Gresham Ship

1595, for example, three London merchants were excused customs duties on 'merchandise brought by the said ships' up to the value of 636 crowns (Oppenheim, 1896, 168). According to the listings for Royal Bounty (Table 5.1), fifteen vessels of 100–160 tons were built between 1574 and 1577 in London and the east coast ports (Dietz, 1991).

Any one of the vessels of an appropriate tonnage listed in Table 5.1 might be the Gresham Ship; all were certainly her contemporaries in date and design. It should be noted, however, that the calculation of tonnage shown here was not an exact science and does not easily equate with modern calculations of net or gross tonnage, rather with capacity, as reflected by the notional number of tuns of Bordeaux wine such a vessel might accommodate. It was not until some years later in 1582 that Matthew Baker, the prominent Elizabethan shipwright, devised his rule for calculating a ship's tonnage with a mathematical formula. The initial example he used to illustrate his methodology was a London merchantman, the *Ascension*, which had a 54 ft (16.5 m) keel, a beam of 24 ft (7.3 m) and a hold 12 ft (3.7 m) deep. She could carry 'in burthen' 160 tons, but her tonnage was calculated at 213½, a figure which was one third higher (Oppenheim, 1896, 132; 71–3). Unfortunately, the Royal Bounty lists do not include every vessel built and regrettably full and reliable histories for all those that are listed cannot be compiled, not least because ships changed not just their spelling, but also their names from time to time, while several vessels could share the same nomenclature; the London Port Book for 1568 shows that *Bonaventure, Christopher, Elizabeth, Falcon, Jesus, James, John, Mary Pelican* and *Spreadeagle* were among the most commonly attributed ship names (Dietz, 1972, 190–5).

What can be inferred from her size, however, is that the Gresham Ship was intended for long distance trade. Brian Dietz (1986, 128) has shown that vessels engaged in London's admittedly lucrative cross-channel trade with north-west Europe were half the size of the vessel usually employed on the longer hauls. The average for such vessels in 1568 is 40 to 80 tons burthen, applying Baker's Rule, or 53 to 106 tonnage. Thus the Gresham Ship represents a particular English vessel type from a particular period in which the development of trade and exploration were as entwined as the political and religious struggles. Consequently the contemporary shipping could be engaged in trade, exploration, piracy or naval warfare. But all these activities were rarely mutually exclusive in the serious competition that ensued with England's maritime rivals, be they the Catholic Spanish and Portuguese or subsequently the Protestant Dutch.

New Worlds

For English mariners, the horizons were widening in the 15th and early 16th century. It was the ships of fishermen, rather than merchantmen, that may have served as the pioneers in this field; they were certainly working, for example, as far north as Iceland by 1415 (Kowaleski, 2000, 29–32; Jackson, 2000, 46–9). In more southerly waters English merchants were also sailing further, trading deep into the Mediterranean, in spite of Italian opposition. In 1446, for example, we hear that goods from Genoese merchants in London were seized and a fine of £6,000 was levied in compensation for an unprovoked attack on an English ship trading in Genoa. Such voyages became increasing common, as this was also the period that saw the establishment of the Merchant Adventurers Company in 1486, through which English merchants exported goods directly to ports overseas.

Rather more daring enterprises are recorded towards the close of the century, with the battle to corner the most lucrative of prize of all, the trade with India and the East. For centuries this had been monopolized by Venice, sitting at the Mediterranean end of land routes to Asia. With the fall of Constantinople in 1453 to the Ottoman Turks, the need to open up a direct maritime route became increasing crucial. This was the imperative that drove the Spanish expedition led by Christopher Columbus to the West Indies in 1492; the round trip took 224 days (Scammell, 1981, 302–5). By this date, the Portuguese had already worked their way round the western coast of Africa and Bartholomew Dias reached the Indian Ocean in 1488; just ten years later Vasco da Gama has sailed to India, then to the Spice Islands in 1512 and finally to China itself by 1513 (Scammell, 1981, 225–43; Afonso *et al.*, 1998,

236–7). It was left to Ferdinand Magellan to achieve the first circumnavigation of the world in 1522, joining the Atlantic and Pacific Oceans. There was a cost involved in such enterprises; broadly contemporary wrecks of vessels built in the Iberian-Atlantic tradition have been discovered off the coast of Texas (San Esteban), the Bahamas (Highborn Cay and Western Ledge), Turks and Caicos Islands (Molasses Reef), Labrador (Red Bay), Kenya (Mombasa) and the Philippines (San Diego) (Afonso *et al.*, 1998, 78–81).

The English would learn from those Spanish and Portuguese pioneers, but their own contribution to this golden age of exploration had begun inauspiciously in the previous century. On 15 July 1480 a Bristol ship captained by John Jay the younger set off westwards on an expedition to the legendary 'Island of Brazil'. Although unsuccessful on this occasion, Bristol did become the port from which Englishmen first reached the New World, sailing under John Cabot in 1497 (Scammell, 1981, 461–2). Merchants from London, as well as Bristol, supported some of Cabot's endeavours and 20 years later a scientific expedition set out from London heading westwards for the Americas, led by one John Rastell. Unfortunately his crew preferred the more immediate rewards of carrying cargoes to Bordeaux in preference to the glory of discovering a New World; they marooned Rastell in Ireland and sailed south to France (Moorhouse, 2005, 170).

Four years later, in 1521, not even the King of England could persuade the London merchants to back a scheme to develop Cabot's 'New Island'. The next attempt was not until 1527, when Henry VIII commissioned an English merchantman, built for the Bordeaux wine trade, for a voyage of discovery; she reached the North American coast, worked her way down to the Caribbean and returned safely home again (Scammell, 1981, 462). But still the capital's gaze was elsewhere, for London merchants dominated the ever dependable cloth trade with Antwerp as never before. Here was ready tangible profit; the risks of the American adventures were gladly left to the West Coast ports. Nevertheless, the world was expanding; the geography had changed. England was no longer on the very northern edge of the Mediterranean theatre or the eastern edge of the Scandinavian world; it now found itself in a central position on the enlarged world map, looking east towards the old world and west, if hesitantly, towards the new. In 1553, for example, Edward VI granted a charter to the 'Company of Merchant Adventurers ... for the Discovery of Regions, Dominions, Island and Places Unknown' and Sir Hugh Willoughby and Richard Chancellor inaugurated a new generation of exploration with their expedition from Deptford to seek out a northern trade route to China (Willan, 1956,4–7).

Once London's merchants and mariners could content themselves with increasing the profitable cargoes carried to and from the great European entrepot at Antwerp (Van der Wee, 1963, 113–40), set in the heart of the Protestant Netherlands. But with the collapse of this once so dependable trading axis in the 1560s, the English merchants were obliged to move on (Ramsay, 1986, 62). They now had to seek new outlets for their wool, cloth and other lucrative trades and thus the English cloth fleet sailed from London to a new destination, initially to Hamburg in March 1569 (Ramsay, 1986, 116). The collapse of the old order obliged the English to direct greater resources and more energy at other markets that had hitherto attracted less attention. In 1550 most of London's merchants traded with their European neighbours from the Zuiderzee to Marbella in the south. But the period from 1574 to *c.* 1600 (the operative life of the Gresham Ship) saw charters awarded to the French, Spanish, Barbary, Eastland, Levant and Muscovy Companies, and ultimately to the East India Company, the organization that became the most commercially successful of them all.

New Found Lands

This turbulent era saw English ships sailing to the Caribbean via the West African coast, such as the voyages undertaken by John Hawkins between 1555 and 1569, expeditions that involved the buying, capturing or selling of slaves (Froude, 1923, 17–36). These ventures, supported by wealthy merchants and Elizabeth herself, set up what was to become the familiar triangular pattern of the slave trade, notoriously linking Africa, the Caribbean and Europe (Kelsey, 2003, 3–18). Then there were English expeditions to North America, seeking a north-west passage to Cathay. The first was initially funded by the Loks (the London merchant family involved in the African slave trade since 1555) and led by Martin Frobisher, variously described as an adventurer, slaver and pirate. Given the pioneering nature of such expeditions, the size of the fleet was surprisingly modest: the total complement of the crew was but 35, manning a pinnace and two merchantmen, the *Gabriel* and the *Michael*, both newly built in 1574 and thus contemporary with the Gresham Ship. Frobisher's first voyage left Blackwall on 7 June 1576, sighting the coast of Labrador at the end of July, and then reached Baffin Island in Frobisher Bay on 18 August. Although the pinnace was lost in a storm and five crew members were captured by Inuit, Frobisher managed to return to London on 9 October without further mishap (McDermott, 2001).

He had brought with him a 'piece of black stone', which some thought might be a gold-bearing ore. This was sufficient to persuade influential backers, including the Queen herself, to fit out a much larger expedition of 150 men. A former naval warship, the *Ayde*, joined the *Gabriel* and the *Michael*, leaving Blackwall on 27 May 1577, reaching Frobisher Bay on 17 July. Having taken possession of this new country in the name of his Queen and having loaded his fleet with 200 tons of the 'black stones', his fleet left for England on 23 August. His third voyage left in 1578, sailing via Greenland, discovered what is now known as Hudson Bay and unsuccessfully attempted to set up a colony in the new lands of *Meta Incognita* (Baffin Island). The voyage back to England set sail in August 1578 with another large consignment of ore. Alas, it was subsequently shown to be iron pyrites and not gold (Symons, 1999, 401–76). The explorations of the northern waters were followed by John Davis in 1585 and again in 1587, moving through the straits between Greenland and Baffin Island (Scammell, 1981, 473).

Meanwhile, Francis Drake circumnavigated the globe between 1577 and 1580, pillaging Spanish treasure en route (Froude, 1923, 50–69). This remarkable expedition enraged Phillip as much as it delighted Elizabeth. His vessel, the *Golden Hind*, was almost identical is size to the Gresham Ship, demonstrating the robustness of such vessels and the ruthless optimism of their masters (Cummins, 1995, 72–123). Between 1584 and 1587 Walter Raleigh and Richard Grenville undertook four expeditions to explore the coast of North America including Florida, North Carolina and the lands named ''Virginia'' in honour of Queen Elizabeth. Although the first attempts at colonization were abortive (Scammell, 1981, 481–3), it was not until 1607 that a permanent settlement in New England was established at Jamestown (Kelso *et al.*, 1997; Noël Hume, 1994).

Thus the relatively half-hearted approach to 'globalization' seen in the early 16th century was transformed into a powerful Protestant imperative, in which profit and privateering at the expense of Catholic countries such as Spain found backing from London financiers, the landed gentry and the even the Crown. Such work, part-trader, part-raider, required robust armed merchantmen and robust armed mariners; England greatly expanded its tradition of building one and breeding the other.

Privateers

Extending London's mercantile reach in such an age was not without risk; the trade with Spain can stand as an example. From at least 1530 a group of English merchants trading in the Iberian Peninsula in the key port of San Lucar established a 'Brotherhood of St George'. From these origins, the Spanish Company was formed, with a Royal Charter signed in 1577; among honorary members listed in that document were Walsingham and Leicester, as well as Sir Thomas Gresham (Croft, 1973, xiv). In spite of such influential patronage peaceful commercial success was not assured. That very same year Francis Drake, taking a diametrically contrary approach, began his belligerent campaign attacking Spanish shipping in the New World (Cummins, 1995, 72–123). Some praised his exploits and Elizabeth was much enriched by them, but a core of the London merchant class were less impressed. There were serious repercussions for anyone engaged in lawful trading ventures, not least for English merchant shipping harbouring in Spain and Englishmen living there. Delicate negotiations were the order of the day, not alas assisted by the new religious thinking. Some of the merchants working in Spain had Spanish wives and remained Catholic; the newer breed of English entrepreneurs were hardline Protestants, while some preferred to be Catholics abroad and Protestants at home (Croft, 1973, xxi).

But by May 1585 diplomacy was failing; outright war was looming and a full-scale invasion of England was being planned. English merchantmen berthed in Spanish ports were impounded and requisitioned to serve in the Armada and the men were imprisoned to await the attentions of the Inquisition (Croft, 1973, xxviii). Only one ship, the *Primrose* of London escaped to relay the news. In the outcry that ensued the High Court of Admiralty began to issue Letters of Reprisal, which authorized anyone who could show that he had lost goods or ships by the action of the Spanish to recover his losses by force, i.e. to capture shipping and cargoes up to the value of his loss. It was only a matter of time before those seeking genuine compensation were joined by anyone – with or without the Lord Admiral's licence – who had a suitably armed vessel capable of capturing any ship or cargo it could find on the high seas.

Piracy, licensed or not, had long been an unwelcome, but not unusual component of maritime life and not just in time of war (Ramsay, 1986, 89; Eddison, 2013; Milne, 2003, 131–2; 162–3). But now there was a dramatic escalation of such activities, building on the model set by the 'Sea Beggars' in the 1560s, the Protestant merchantmen from the Netherlands, fighting for their independence from their Spanish overlords. A new era of 'privateering', a loosely legalized form of piracy, began in England in earnest. This war with Spain was thus waged not directly by the state itself, but by its merchant seamen, an 'anti-Spanish enterprise at sea' (Spence, 1995, 64). Nevertheless, it was sanctioned by Charles Howard, the Lord High Admiral, and financed by the profits it made (Andrews, 1964, 3–6). London led this particular field and from 1589–91, when some seventy vessels undertook privateering expeditions from that port, capturing 71 prizes valued at £97,283; this was as much as all the sea dogs in Devon, Cornwall, Bristol and Bridgewater combined (Andrews, 1964, 125). To take the year 1590 as an example, Table 5.3 show that 21 recognized privateering expeditions left London, often single vessels, but sometimes two together (six examples), a convoy of three and two fleets of five each. Most were London vessels, but some linked up with ships from other ports such as Bristol, Dover, Plymouth and Southampton.

Something of the organization behind these enterprises is recorded in Table 5.2 in which the promoters who provided the financial backing are listed. The names of London merchants, singly or in consortia, are often

London Privateers 1589–1591

Ship	Tonnage	Captain	Promoter(s)	Date
Adonia	?	?	R Sadler, N Clevinger,	1589
Amity	100 tons	Thomas White	H Colthurst & Co	1589
Amity	100 tons	Walter Crunnelowe	H Colthurst & Co	1590
Amity	100 tons	Thomas White	H Colthurst & Co	1591
Balinus	40 tons	John White	G Bassett	1591
Black Dog	60 tons	William Michelson	Wm Michelson, H Cletherow	1588–9
Black Falcon	?	Thomas Foscue	Ed Glenham	1591
Black Burr	130 tons	William Irish	Sir Geo Carey	1591
Centaur	120 tons	William Lane	J Watts *et al.*	1591
Centurion	200 tons	Christopher Firth	City of London	1591
Cherabin	250 tons	Brooters	City of London	1591
Costly	200 tons	?	City of London	1591
Crocodile	?	George Deane	Ed Fenner	1591
Delight	40 tons	John Thomas	J Barber, T Matthew	1590
Delight	40 tons	John Ridleaden	J Barber, T Matthew	1591
Delight	?	Thomas Beste	T Bramley, R Ofield	1590
Discharge	50 tons	Robert Hutton	Henry Seckford	1589
Discharge	50 tons	Thomas West	John Watts	1590
Discharge	50 tons	Thomas West	T Fenner, J Bird	1591
Dolphin	90 tons	John Maynard	H Colthurst & Co	1589
Dolphin	50 tons	?	John Watts	1590
Eagle	?	Henry White	H Colthurst & Co	1589
Eagle	?	?	H Colthurst & Co	1590
Ed Bonaventure	250 tons	James Lancaster	Thos Cordwell	1590
Ed Bonaventure	250 tons	James Lancaster	Thos Cordwell	1591
Edward Constance	?	Edward Glenham	Ed Glenham	1591
Elizabeth	120 tons	John Matthews	W Hall, J Howell, W Russell	1591
Elizabeth & Mary	?	William Myddelton	Thos Myddelton	1591
Examiner	?	Anthony Barlowe	J Watts, Thos Sewell	1590
Fancy	60 tons	Nicholas Webb	A Fones, E Ansell, P Darrell	1590
Fancy	60 tons	Nicholas Webb	A Fones, E Ansell, P Darrell	1591
Fortunatus	40 tons	George Martin	?	1590
Globe	120 tons	Nicholas Nelson	Ed Leckland, R Sadler	1591
Golden Dragon	150 tons	Roger Roidon		
Golden Noble	200 tons	Edward Partridge	J Bird, J Newton	1591
Grace of God	35t	William Grafton	Wm Grafton	1591
Greyhound	?	Christopher Colthurst	?	1591
Bark Hall	60 tons	John Langford	Wm Hall	1591
Handmaid	?	?	City of London	1591
Hare	?	Geede	Thos Mydelton, E Harby	1591
Harry & John	150 tons	Abraham Cocke	J Watts, J Stokes, H Cletherow *et al.*	1590
Harry & John	150 tons	Wm Craston	J Watts *et al.*	1591
Harte	40 tons	Thos Mydelton	T Mydelton	1591
(Little) John	100 tons	John Musset	J Watts, J Stokes	1589
(Little) John	100 tons	Christopher Newport	J Watts, J Stokes	1590
(Little) John	100 tons	Michael Geare	J Watts, J Stokes	1591
John Evangelist	?	Wm Lane	J Watts	1590
Lark	40 tons	?	H Colthurst & Co	1590

Table 5.2 London ships, masters and promoters recorded by the High Court of Admiralty as having undertaken privateering common and regularized occupation in London

Ship	Tonnage	Captain	Promoter(s)	Date
Magdalen	*50 tons*	*?*	*City of London*	*1591*
Margaret	60 tons	Robert Hallett	R Cobb, J More, R Southwick	1589
Margaret & John	180 tons	John Rickman	J Watts	1591
Mary	60 tons	Edmund Barker	J Jackson, A Jackson, E Arpnsell, J Lancaster	1590
Mary	60 tons	Allan Coates	J Jackson, A Jackson, E Arpnsell, J Lancaster	1591
Mary	?	Arthur Strecke	M Strecke	1590
Mayflower	300 tons	?	?	1591
May Morning	50 tons	Abraham Crocke	J Watts	1589
Merchant Royal	300 tons	Samuel Foxcroft	T Cordell	1590
Merchant Royal	300 tons	Samuel Foxcroft	T Cordell	1591
Mermaid	100 tons	Thomas Huckley	Wm Bond	1591
Moonlight	80 tons	Edward Spicer	Wm Sanderson	1590
Moonshine	50 tons	John Myddelton	J Newton, J Bird	1590
Moonshine	50 tons	John Myddelton	J Newton, J Bird	1591
Passport	80 tons	Christopher Colthurst	H Cothurst & Co	1589
Passport	80 tons	Oliver Maister	H Cothurst & Co	1590
Passport	80 tons	Thomas Rise	H Cothurst & Co	1591
Pegasus	80 tons	Stephen Michel	J Watts	1591
Penelope	?	George Raymond	G Raymond	1590
Penelope	?	George Raymond	G Raymond	1591
Post	30 tons	?	R Cobb, T Pullison, T Starkey	1590
Pretence	60 tons	John Thomas	S Thomas	1591
Prudence	70 tons	John Brough	R Cobb, J Moore, W Jones, J Newton	1591
Return	?	Thomas Philips	Henry Seckford	1589
Return	?	Thomas Philips	Henry Seckford	1590
Robert	?	Nicholas Downton	Earl of Cumberland	1590
Roebuck	240 tons	John Cocke	T Cavendish	1590
Roebuck	240 tons	John Cocke	T Cavendish	1591
Salamander	90 tons	Thomas Ellis	Wm Holliday	1589
Salamander	90 tons	Anthony Potts	Wm Holliday	1590
Samaritan	160 tons	William Jones	T Bramley, R Ofield	1590
Seadragon	50 tons	Robert Naylor	W Holliday	1589
Seadragon	50 tons	Anthony Ingram	W Holliday	1590
Seadragon	50 tons	John Lagatt	W Holliday	1591
Squirrel	40 tons	?:	?	1591
Stephen	180 tons	Thomas Horwell	A Fursland, J Cherry	1590
Stephen	180 tons	?	A Fursland, J Cherry	1591
Supply	?	Peter Lee	Henry Seckford	1589
Susan	300 tons	Robert Flick	City of London	1591
Swallow	70 tons	George Raymond	G Rainbridge, T Stepney	1590
Swallow	70 tons	John Johnson	G Rainbridge, T Stepney	1591
Swiftsure	80 tons	James Beare	G Raymond	1589
Tiger	170 tons	John Markham	W Holiday, J Watkinson, A Fursland, R Bowles	1590
Tiger	170 tons	George Barnestraw	W Holiday, R Bowles, G Barnestraw, M Bank	1591
White Lion	340 tons	Thomas Polwhele	J Chidley *et al.*	1589
Wildman	300 tons	John Chidley	J Chidley *et al.*	1589
Wildman's Club	25t	Richard Glover	J Chidley *et al.*	1589

expeditions 1588–1591 (after Andrews, 1964: 243–9). Note how many names are repeated, demonstrating that privateering was a

London Privateer Expeditions 1588–1591			
Year	Ship(s)	Ship(s)	Ship(s)
1588–9	*Adonia* with *Grace of God* (Topsham)	*Black Dog*	
1589	*Amity with Dolphin*	*Discharge* with *Return*	*Dolphin* with *May Morning*
	Eagle	*Elizabeth & Mary*	*(Little) John* with *Minion* (Bridgewater)
	Margaret with *Supply*	*Passport*	*Salamander*
	Seadragon	*Swiftsure* with *Julian* (Lyme), *Delight* (Lyme) *Unicorn* (Barnstaple)	*White Lion* with *Wildman, Wildman's Club* and *Robin* (Bristol)
1590	*Amity* with *Eagle, Lark, Passport, Post*	*Delight*	*Delight* with *Samaritan*
	Edward Bonaventure with *Merchant Royal*	*Examiner*	*Fancy* with *Grace of God* (Dover)
	Fortunatus	*Harry & John* with *Little John, John Evangelist, Moonlight, Conclude*	*Mary* (or *Bark Smith)* with *Delight* (Southampton)
	Mary	*Moonshine*	*Penelope* with *Pilgrim*
	Passport	*Return*	*Robert*
	Roebuck	*Salamander*	*Samaritan*
	Seadragon	*Stephen* with *Tiger, Minion* (Bristol)	*Swallow* with *Dainty* (Plymouth)
1591	*Amity*	*Balinus*	*Black Falcon* with *Edward Constance*
	Black Burr with *Content, Swallow* (Southampton)	*Centaur* with *Harry & John, Little John, Pegasus, Fifth Part, Bark Burr, Content, Swallow*	*Centurion* with *Cherubim, Costly, Handmaid, Magdalen, Mayflower, Margaret & John,*
	Squirrel, Susan	*Crocodile*	*Delight*
	Discharge with *Golden Noble, Falcon* (Portsmouth), *Elizabeth* (Portsmouth)	*Edward Bonaventure* with *Merchant Royal, Penelope*	*Elizabeth*
	Fancy (wrecked)	*Globe* with *Swallow, Bark Bond* (Weymouth)	*Golden Dragon*
	Grace of God	*Greyhound*	*Bark Hall*
	Harte	*Margaret* with *Prudence*	*Mary* (or *Bark Smith)*
	Mermaid	*Moonshine*	*Passport*
	Pretence	*Roebuck*	*Seadragon*
	Tiger with *White Lion* (Weymouth)		

Table 5.3 *London ships recorded by the High Court of Admiralty as having undertaken privateering expeditions between 1588 and 1591; some vessels sailed singly, others in groups, sometimes with ships from other ports (after Andrews, 1964, 243–9). This was big business: in 1591, for example, there were fifty London ships thus engaged*

repeated, showing that, for the likes of John Watts, John Stokes, Thomas Cordell and Henry Cothurst, privateering was being run as a business. Ships and crew were hired and were often sent off on annual expeditions, year after year (see for example the *Amity*, the *Discharge*, the *Little John*, the *Seadragon*). The range of ship sizes is perhaps unexpected, for the tonnage of some is recorded as between 25 and 40 (the *Wildman's Club*, the *Post*, the *Squirrel*), rather modest vessels, alongside the 200-ton *Centurion*, the 250-ton *Cherabin* and the *Edward Bonaventure*, and the giant 300-ton *Merchant Royal*. By no mean all headed for West Indies, as during this period of maritime anarchy, any Spanish, French or Portuguese prize was worth capturing, wherever it was found. Nevertheless, the Caribbean provides perhaps the best known theatre in which this unorthodox mercantile strategy was played out. Spain counted the cost of holding onto its mighty maritime empire, witnessing the effects of some seventy English naval expeditions between 1585 and 1604, in which it is suggested that a staggering 1000 Spanish and Portuguese vessels were captured. English privateering was big business in these decades, significantly contributing to the Gross Domestic Product (Scammell, 1981, 471).

English ships were especially fitted out and armed for such work, with 'suitable' captains and crew engaged. Such expeditions required cash to set up and this was provided by prominent individuals, such as the Earl of Cumberland (Spence, 1995. 64–83) or by a group of merchants, such as Bayning, Boreman, Cordell, Flick, Lee and Watts, the consortium behind the audacious plundering of Cadiz harbour in 1587 (Andrews, 1964, 94; Mattingly, 1959, 100–14). As for the captains of such privateering ships, the biographies of Michael Geare and Christopher Newport, two mariners who both lived in London, will stand as examples. Geare made his first two reprisal voyages in 1586/7 in the *Marlyn*, starting as 19-year old crewman, then in the West Indies with the *Drake* (1588), the *Examiner* (1589), the *Little John* (1590, 1591), a ship he then commanded as captain. He made sufficient money from his highly profitable operations to become part-owner of that ship with his new backer, John Stokes; the ship was then renamed the *Michael & John.* Four more successful voyages in her followed, after which he worked with Christopher Newport in 1596 and 1602–3 (see below), as well as with David Myddleton (1601–2). He was knighted and thus died rich and respectable in Stepney, an Elder Brother of Trinity House. But since his will saw the distribution of five pounds annually among the poorest decayed seaman or their widows in the Thamesside hamlet of Limehouse (Andrews, 1964, 87–8), he had not forgotten his humble origins.

Christopher Newport offers another example of a mariner's rise through the ranks of Elizabethan society. He is first recorded in 1580 when, aged 20, he sailed in the *Minion* of London in one of the first English expeditions to Brazil. By 1582, he was working on ships from Harwich, but by 1587 he was the Master's Mate on the *Drake*, owned by London merchant John Watts, one of the fleet that attacked Cadiz in that year, before prize-hunting along the rest of the Spanish coast. By 1589, he was master of the privateer *Margaret* of London and was then sent by Watts to the West Indies in the *Little John*, which he then commanded. In a fierce fight with a Mexican ship off Cuba, his right arm was 'strooken off', but that did not deter him from four further voyages to the Caribbean for London merchants. His ship was now the *Golden Dragon*, which he took on a legitimate trading venture through the Mediterranean, although 'Captain One Hand' returned with two Spanish prizes in tow. From 1596, as captain of the *Neptune*, described as a formidable man-of-war, he led aggressive expeditions to the West Indies, returning with the very substantial loot from the ships and towns he ferociously pillaged. His fearsome reputation saw him chosen by the Virginia Company in 1606 as the Admiral of their fleet for the first six years of the fledgling New England colony. The town of Newport, Kentucky, was named after him in recognition of the valuable work he did for the colonists. He did not retire peacefully to enjoy his considerable wealth, as might have been expected. Instead he turned his attentions to the other side of the known world, undertaking three voyages for the burgeoning East India Company, helping to lay secure foundations for their phenomenal success. He was by no means the only privateer to turn his gaze eastwards; the Earl of Cumberland, for example, who was involved in attacks on Puerto Rico in 1597–8, was an active participant in the new EIC, for it was his ship, the *Red Dragon*, that left Woolwich on 13 February 1601 on the company's first formal trading expedition (Spence, 1995, 77–8). As for Christopher Newport, he died in the service of the company at Bantam, Java, in 1617, aged 57. For a man described in 1584 as 'of Limehouse, a mariner', his life encapsulates a truly remarkable career as seaman, privateer, colonist merchant and adventurer (Andrews, 1964, 84–6). Such men, such ships.

Levanters

Study of the Turkey and Levant Company records, together with other contemporary sources, also provides an economic, political and a social context for late 16th-or early 17th-century merchant shipping such as the Gresham Ship. The first Charter of the Turkey Company dates to 1581 and the first charter of the Venice Company to 1583, which finally led to the founding of the Levant Company in 1592. A review of the company's activities from about 1585 to 1590 notes that their fleet initially comprised some nineteen ships, which between them made twenty-seven voyages, using a combined crew of 787 men (an average of about forty per ship). The customs duties alone paid over this period was some £11,359 6s, suggesting that the trade was extensive and presumably very profitable (Epstein, 1908, 232–8).

Study of the company records together with other contemporary sources provides an economic, political and social background to the life of the crews of late 16th-century merchantmen such as the Gresham Ship. The cargoes carried usually included cloth (mostly purple or red), but other key commodities included tin, lead and furs. Impositions charged on 87 categories of goods imported into Turkey since at least the early 17th century included iron at 2 shillings per ton, lead at 1s 6d per fother (= 19½ cwt/990 kg), red lead at 3d per hundred (51 kg) and tin at 1s per hundred (Epstein, 1908, 232–6); the cargo carried by the Gresham ship would not have been out of place here. It is perhaps of some significance to note that, in a letter illustrating typical charges levied on the Levant Company, the two commodities exported in bulk used as examples were kersies (cloth) and tin, the latter purchased by the 10 thousand-weight (508 kg) (Epstein, 1908, 262–3). Certainly the London customs record for 1560 shows that iron was a major commodity there (presumably brought in for re-export); it was ranked in the top ten in terms of cost, below linen, fustians, canvas, woad and various wines (Dietz, 1972, 152–3). Looking further afield, by the 1580s English merchantman were trading not just length of the north European coast, but right around the Mediterranean including North Africa, also to the Atlantic Islands and then to Central and South America. Although the imports they returned with varied as much as the ports themselves, the exports they carried out invariably included staples such as cloth, lead tin and hides (Bindoff, 1950, 285).

In the 'Lists of Ships Belonging to the Levant Company' (Epstein, 1908, 217–29), fourteen vessels are named as trading with Zante (Zakinthos) and Candia (Crete) in 1581. These include the *Royal Merchant,* described as a vessel of 350 tons burthen with a crew of 60 men, statistics that are repeated in the ship list for 1584. Although clearly larger than the Gresham Ship, its story is worth recounting, not least because both vessels were built in 1574. The names of many of the company ships are the same, from year to year (see Table 5.4). The operative life of a hard-working trader in the late 16th century could be considerable; Ian Friel's study of records made for the High Court of Admiralty suggests that, of the English vessels assessed, some 70% were between 11 and 50 years old (Friel 1995, 188). Its seems that each Levanter undertook one major voyage each year, usually sailing from London, around the Iberian coast, through the Straits of Gibraltar and then heading for Venice, Constantinople or Scanderoon (Tripoli, Lebanon), via a series of other entrepots. The regularity of the voyages and repetition of the ship names suggests that the many of the crews may have also been retained. The larger ships had a crew of 50 to 60 on board for the duration of the annual voyage, arguably as testing a social experience as it was physically demanding. But ships often travelled in pairs or in convoy and the voyages were punctuated by periods on shore, as cargoes were unloaded and new cargoes collected. The crews from all the company ships would thus have the opportunity to meet up with each other at regular intervals on *terra firma*, a welcome antidote to cabin fever, as well as affording opportunities to see new lands, learn new languages and appraise different cultures. This was a period in which England's mariners enjoyed far wider horizons than the majority of the population in their native islands.

The dangers faced on such ventures are described in the second volume of Hakluyt's *Principal Navigations*, in his account of sea battles in the Mediterranean. Five of the Company's ships had left London in November 1585; the *Merchant Royal* and the *William and John* were bound for Tripoli, the *Toby* for Constantinople and the *Susan* and the *Edward Bonaventure* for Venice. Having discharged their exports and taken on their return cargoes, they all headed for the rendezvous at Zante before setting out for England in convoy. On 13 July 1586 they were attacked by Spanish and Maltese galleys between Tunis and Sicily, near the island of Pantelleria, but beat them off. The English ships then put into Algiers for supplies, before fending off a second group of Spanish galleys near Gibraltar (Hakluyt, 1599, 285–9; Edward de Vere Newsletter, 1989; 2001, **2**, 1).

As for the *Merchant Royal*, it seems to have returned to its usual business, since it is listed in the catalogue of nineteen vessels 'set to work by the Turkish Company' between 1583 and 1588 (Table 5.4). But for at least two of those years it found employment elsewhere, having been requisitioned for war work against Spain's 'Great Enterprise', the attempt to invade Protestant England. In 1587, for example, it was fighting alongside Francis Drake at Cadiz, singeing the beard of the Catholic King Phillip of Spain. This was an extraordinary pre-emptive strike against the Great Enterprise. The resulting delay bought precious time and allowed Elizabeth to be better prepared to face the threat (Mattingly, 1959, 100–14).

Then in August 1588 Phillip finally launched his massive armada. The plan was for the huge fleet to sail from Spain, heading east along the southern coast of England. This formidable naval force would then rendezvous with a fleet of troop transports, which would require protection during their short sea crossing from Calais to Kent, where the planned military invasion would begin. Unless the Spanish Armada could be beaten or the rendezvous disrupted, England would be invaded, just as the Protestant Netherlands had been. The Queen's Navy comprised just 34 ships and consequently many of the country's armed merchantmen were quickly pressed into service. The City of London, by far the richest port in the land, readily supplied a further 2130 men and over 30 ships. Having already tasted naval warfare under Drake's command the previous year, the London ships formed a squadron of some consequence and acquitted themselves honourably in the tense weeks that followed. It was a close run thing and, although unsettled weather and storms – God's Breath – ultimately rescued the Protestant cause, there can be no doubt that the presence, seafaring skills and discipline of the large numbers of armed merchantmen that provided the backbone of the English fleet were positive factors in the eventual victory.

In addition to the two redoubtable Levanters, the *Merchant Royal* (400 tons) and the *Edward Bonaventure* (300 tons), the London Squadron that faced the Spanish Armada is listed in Table 5.5. Again, the Gresham Ship might be one of these; almost certainly some of those who sailed in her during her long life must have been gainfully employed in this operation. It is worth emphasizing that close bonding between the crews of these merchant ships must have been provided by the shared dangers of the job; in 1585, as we have already noted, a convoy of five ships of the Levant Company came under attack by Spanish and Maltese galleys, but working together, the merchantmen saw off the aggressors. Three of those same ships that shared that experience were

Merchantmen serving with the London Squadron 1588

400 tons	*Merchant Royal*	160 tons	*Bark Bur*
300 tons	*Edward Bonaventure*	160 tons	*Brave*
300 tons	*Hercules*	160 tons	*Royal Defence*
250 tons	*Toby*	160 tons	*Vineyard*
250 tons	*Centurion*	140 tons	*Golden Lion*
220 tons	*Violet*	140 tons	*Thomas Bonaventure*
220 tons	*Susan Parnell*	140 tons	*Samuel*
200 tons	*Mayflower*	120 tons	*George Noble*
200 tons	*Minion*	120 tons	*Toby*
200 tons	*Ascension*	120 tons	*Antelope*
200 tons	*Primrose*	100 tons	*Anthony*
200 tons	*Margaret & John*	100 tons	*Pansy*
200 tons	*Tiger*	100 tons	*Jane Bonaventure*
200 tons	*Red Lion*	80 tons	*Diana*
200 tons	*George Bonaventure*	80 tons	*Passport*
180 tons	*Gift of God*	60 tons	*Moonshine*
180 tons	*Anne Frances*	60 tons	*Release*
170 tons	*Solomon*		

Merchant vessels known to have been used as fire ships 1588

200 tons	*Bark Talbot*	?	150 tons	*Cure's Ship*	?
200 tons	*Thomas Drake*	?	140 tons	*Bear Young*	?
180 tons	*Hope*	Plymouth	120 tons	*Angel*	Southampton
150 tons	*Bark Bond*	?	90 tons	*Elizabeth*	Lowestoft

Table 5.5 *London merchantmen requisitioned for active service with the London Squadron against the Spanish Armada in 1588, listed in order of size (tonnage) (after Laughton, 1981, 324–31; National Maritime Museum, 1988, 156–8). The Gresham Ship might be one of these vessels: almost certainly some of those who sailed in her during her long life must have been gainfully employed in this operation.*

subsequently called upon to join Francis Drake in the attack on Cadiz harbour in 1587, while at least fourteen of the Levant Company's ships also served alongside the Queen's navy to take on the might of the Spanish Armada the following year. The battle-hardened crews then returned to their 'normal' duties of the annual voyage from London to Constantinople, but there must have been an increased bond and mutual respect felt by the men of such vessels as the *Merchant Royal* and the *Edward Bonaventure*, given their joint actions in the Mediterranean, at Cadiz and in the Narrow Seas.

That said, the life at sea remained hard, with death by accident or disease more common than by enemy action. Consequently, most ships would need a doctor. One is specifically mentioned in the records of the High Court of Admiralty in the crew of a 14-gun, 100-ton vessel called the *Elizabeth*, which was thus not of dissimilar size to the Gresham Ship. His name was Thomas Baker Junior, for he was the son of the Commander, who was also one of the eight owners of the ship. Presumably whatever profits a voyage made would be subdivided at least eight ways. The full complement was sixty, but only the names of the more senior crew members are recorded: the Master, John Hubbard; Lieutenant, Elisha Pobree; Boatswain, John Dolby; Gunner, Philip Cartwright; Carpenter, John Limbrey and the Cook, Richard Point d'Estre (PRO, High Court of Admiralty, HCA, 26/3/59).

We learn more about the crews of such vessels when such ships were requisitioned for military duty. The wages for the rank-and-file seaman, when in the service of the crown, rose from 6s 8d a month in 1546 to 10s per month in 1585 (Oppenheim 1896, 134). The Gresham Ship would accord with what could be termed a fourth-rate in the 16th century, i.e. vessel of between 100 and 200 tons (Oppenheim, 1896, 118). Table 5.6 shows the monthly pay for the crew of such a ship in 1582. It exhibits some interesting priorities (the gunner is paid less than the cook) and highlights the importance of, for example, the trumpeter.

To return to the merchantmen plying the Levant trade: the name of the *Royal Merchant*, described as of 350 tons burthen and with a crew of 70 men, appears in a list of 'ships & mariners maintained and set to work by the merchants trading Turkey & the Dominions of the Seigniory of Venice', dated 20 May 1590 (Epstein, 1908, 217–29). Although normal service might seem to have been resumed following the Spanish emergency, the Levant Company's ship-lists show no further mention of a ship called the *Royal Merchant* or indeed the *Merchant Royal* until 1600, although some thirty-

Names of Ships Working for the Levant Company 1581–1600

Epstein, 1908, App. V	1 1581	2 1584	3 1583–8	4 1590	5&6 15??	7 15??	8 1595	9 1600
Alice Thomas		140 tons/28						
Alice Thomas					-tons/30			
Angel						200 tons/50		
Angel							180 tons/40	
Angel								140 tons/35
Anne Frane								140 tons/35
Ascension	180 tons/35							
Ascension		170 tons/36						
Ascension				200/38				
Ascension					-tons/35			
Ascension						300 tons/60		
Ascension							300 tons/60	
Alice & Thomas				140 tons/34				
Bark Burre (Bear?)			120 tons/30					
Barque Reynolds			150 tons/40					
Centurion				200 tons/44		200 tons/50	200 tons/44	
Charity	140 tons/30							
Charity			130 tons/36			100 tons/40		160 tons/36
Cherubim				220 tons/44		200 tons/50		220 tons/50
Christ	100 tons/22							
Consent						300 tons/60		
Darling								240 tons/50
Desire								120 tons/30
Diamond								140 tons/35
Dragon								160 tons/40
Eagle				150 tons/38				
Edward Bonadventure	250 tons/50	240 tons/40		250 tons/50				
Edward Bonadventure								160 tons/40
Elizabeth Cocken			120 tons/30					
Elizabeth Stoaks			120 tons/30					
Elnathan								80 tons/20
Emmanuel			200 tons/50					
George Bonadventure	150 tons/35	160 tons/34		180 tons/37	-tons/30			
George Bonadventure						100 tons/40	140 tons/35	
Gift of God	120 tons/24							
Gift of God		130 tons/27						
Gift of God			120 tons/30					
Gift of God					-tons/25			
Golden Noble			140 tons/40					
Great Suzanne						350 tons/70		
Great Suzanne							260 tons/66	300 tons/60
Greyhound								140 tons/35
Guest								120 tons/35
Gyllyon (?Julian)							130 tons/30	
Harry	120 tons/26							
Harry Bonadventure		120 tons/27						
Hector								350 tons/70
Hercules			300 tons/70	300 tons/60		300 tons/60	250 tons/60	
Jesus			100 tons/25					
Jewel							130 tons/30	
John Francis								130 tons/35

Table 5.4 *London ships mentioned in nine separate documents dating from 1581–1600, as recorded by the Levant Company,* Merchant *is also known as* Merchant Royal. *A ship of that name served the company for some 20 years*

	1581	1584	1583–8	1590	15??	15??	1595	1600
Jollian (?Julian)						100 tons/40		
Lanavit(?)								100 tons/25
Lion					140 tons/-			
Little George								100 tons/25
Margaret Bonadventure		120 tons/26			-tons/28			
Mary Martin			160 tons/40					
Mary Rose		130 tons/28	130 tons/40	140 tons/34	-tons/30			
Mayflower			160 tons/40	180 tons/38				
Mayflower						300 tons/60		
Merchant Bonadventure						100 tons/40		
Merchant Bonadventure							150 tons/35	
Merchant Royall			300 tons/70					
Mignon				200 tons/44		200 tons/50		
Margaret & John				180 tons/40				
Phoenix					-tons/25			
Phoenix								300 tons/50
Primrose	300 tons/60							
Prosperous							200 tons/44	
Providence	200 tons/40							
Rebecca								160 tons/36
Recovery							80 tons/20	
Red Lion		150 tons/30		160 tons/38				
Report						200 tons/50		
Royal Exchange						350 tons/70		
Royal Exchange							250 tons/66	
Royal Merchant	350 tons/60	350 tons/60						
Royal Merchant				350 tons/70				
Royal Merchant								300 tons/60
Saker								140 tons/36
Samaritan		140 tons/30						
Samaritan				140 tons/34				
Samaritan					120 tons/-			
Solomon			130 tons/36					
Solomon				180 tons/40				
Suzanne	300 tons/50							
Suzanne		260 tons/50	250 tons/60					
Suzanne				300 tons/60				
Suzanne					200 tons/-			
Suzanne Parnell								200 tons/45
Swallow								150 tons/35
Teagre (?)			150 tons/40					
Thomas Bonadventure	100 tons/25	100 tons/25						
Thomas Bonadventure				120 tons/28				
Thomas Bonadventure					-tons/24			
Toby	140 tons/28							
Toby			200 tons/50					
Toby				230 tons/46				
Trinity Bear		140 tons/30						
White Hind	100 tons/25							
William Fortune								200 tons/45
William & John			120 tons/30					
The ? & Elizabeth							60 tons/16	

showing tonnage and crew numbers. The tonnage listed varies, as does the spelling/naming of the vessels; for example, the Royal *(after Epstein, 1908, Appendix V)*

eight other vessels are named during this period (Table 5.4). Other sources help explain its temporary absence from the Mediterranean, at least for 1591–2, for it was battling around the Cape of Good Hope.

In April 1591, the first English 'trading' voyage to the East Indies set sail from London under the command of Captain James Lancaster and George Raymond. The three vessels were those two old warriors, the *Edward Bonaventure* and the *Merchant Royal*, along with the *Penelope*. Financial backing came from various sources, including, apparently, the proceeds of previous privateering ventures (Scammell, 1981, 475). Not all went according to plan, however. After rounding the Cape, so many of the crew were sick that the *Merchant Royal* was obliged to return to England and then the Penelope was wrecked in a storm at sea. Undeterred, the *Edward Bonaventure* battled bravely onwards to the East Indies. The return voyage was tortuous; after rounding the Cape in March 1593, contrary winds and incipient mutiny took the ship to the West Indies, where it was hit by a hurricane off Bermuda. The *Edward Bonaventure* eventually returned to England in 1594. Although the long voyage's financial success is debated (Foster, 1933, xii–xvii), its significance for the establishment of the East India Company just six years later is not in dispute.

Another brief mention of the *Edward Bonaventure* occurs in 1599, when Thomas Dallam, who travelled to the Levant in that year to present an organ to Sultan Mahomed III as a gift from Queen Elizabeth, mentions in his diary that he left Zante in the *Heckter* on 26 February in the company of the *Edward Bonaventure* and the *Swallow*. We last hear of the *Edward Bonaventure* in 1602, when John Sanderson records that, in February of that year, he took passage in that ship from Scanderoon to Zante. The manner of Sanderson's reference to his voyage in the *Edward Bonaventure* indicates that the vessel was still sailing as a merchant ship for the Turkey Company. Sanderson then notes in his diary that the *Edward Bonaventure* was 'wrecked the year after upon the coast of France, before Boulogne' (Epstein, 1908, 127–28).

Wrecked in the Thames Estuary

Whatever the century, all ships faced the danger of shipwreck on any voyage, given the uncertainties of the weather and of coastal topography. Perhaps the most telling example of such perils is the tale of the *Sea Venture*, which, battered by a hurricane, was thrown onto a reef off Bermuda, but the whole company was ferried to safety on the island. The tale is immortalized as the core of Shakespeare's play *The Tempest*, while the wreck site has been fully investigated (Adams, 2003, 109–22). It is of some significance that the bard mentions shipwrecks so frequently in his plays: in *The Comedy of Errors*, *Twelfth Night*, *Pericles*, while the denouement in *The Merchant of Venice* is famously

Monthly wages of the crew serving in a Merchantman in 1582

Master	£1 11s 3d	(=375d)
Steward	17s 6d	(=210d)
Pilot	16s 8d	(=200d)
Surgeon	15s	(=180d)
Trumpeter	15s	(=180d)
Master's Mate	11s 8d	(=140d)
Boatswain	11s 8d	(=140d)
Quartermaster	11s 8d	(=140d)
Purser	11s 8d	(=140d)
Master Carpenter	11s 8d	(=140d)
Cook	11s 8d	(=140d)
Master Gunner	10s	(=120d)
Boatswain's mate	9s 2d	(=110d)
Quartermaster's mate	9s 2d	(=110d)
Carpenter's mate	9s 2d	(=110d)
Gunner's mate	7s 6d	(=90d)

Table 5.6 Monthly ages of the crew serving in a merchantman in 1582. The range is for a vessel of 100–200 tons and the wages are shown in pounds (£) shillings (s) and pence (d). The Master earns four times more than the Gunner's Mate. (Oppenheim, 1896, 152–3)

triggered by the loss of Antonio's ships at sea. It seems that shipwreck loomed ominously large in the Elizabethan imagination, in much the same way as the loss of a jumbo jet resonates today (Gerald Grainge, pers. comm.).

Even the tidal Thames estuary that claimed the Gresham Ship could be a treacherous waterway with its shifting sandbanks, shallows and shoals. The records of the High Court of Admiralty illustrate this all too well, as a 19th-century incident citing the actions of Issac Jarman, the coxswain of the Ramsgate lifeboat shows. On the night of the 3/4 December 1863 during dreadful weather, rocket signals were seen coming from the Tongue Lightship, which indicated a ship in distress. The Ramsgate steam tug *Aid* set out at 8.45 pm with the open Beeching class lifeboat *Northumberland* in tow. In mountainous seas and with a near hurricane raging, they reached the London to Melbourne emigrant ship *Fusilier* fast in the sand on the Girdler Bank some miles north of Herne Bay, Kent. Going alongside at 2 am, the lifeboat took off, one by one, 40 women and children and 36 men. Three journeys had to be undertaken with huge seas sweeping over the open boat to transfer the survivors to the tug waiting nearly a mile away in the deeper waters of the Princes Channel. That operation took until 6 am. The tug and the open lifeboat were then called to another distressed vessel, aground on the Shingles Bank, from which a further 18 men were saved and eventually landed safely at Ramsgate. Such storms are not infrequent in the North Sea, but fortunately so too are such acts of extraordinary bravery

The very real dangers faced by shipping in the Thames Estuary should therefore not be underestimated and consequently neither should the value of the buoys and beacons set out and maintained by the Brethren of Trinity House. These mark the safe channels, as well as the banks to be avoided, thus facilitating a safe passage. Conversely, in time of war, the removal of these markers would form a first line of defence; as several shipmasters stated in 1545:

> If the beacon standing in the sea at the Spanish Knock be taken away that no stranger-born pilot or shipmaster dwelling in an outward realm can dare lead or bring any ship of charge over [the North Foreland] without extreme danger (Harris, 1969, 173).

Such extreme danger is shown by the fate of the *Cherabin*, a merchantman that sank in a violent storm in the Thames. Since a vessel of that name (or a version of it – *Cherabim* or *Cherubim*) is also listed in the records of the Levant Company in 1590 and in 1600 (as well as in an unspecifically dated record in between), she worked as a Levanter for at least part of her life. Her tonnage was noted as being between 200 and 220 tons burthen and her crew also varied from 44 to 50 (Table 5.4) She thus served alongside such stalwarts of the Turkey trade as the *Merchant Royal*, the *Edward Bonaventure* and the *Hercules*, whose exploits have been summarized above. If this was the same *Cherabin* that had been supplied with two sakers in London in 1590, then she was clearly armed. The guns were purchased for her by a merchant called Edward Johnson from the dealer John Chapman, a prominent figure in the London gun trade. The transaction is recorded in the bonds issued that year for guns sold in London to merchantman; interestingly, Chapman was also responsible that year for the sale of six falcons to another Levanter, the 300-ton *Suzanne*.

A use for her armaments is recorded in 1591, when a London ship called the *Cherubim* was working with Robert Flick's 300 tons *Susan* (or *Suzanne*?), part of the London squadron sent to reinforce Lord Thomas Howard's fleet in the Azores. This was a privateering enterprise, in which English ships tried to capture Spanish vessels and their cargoes heading back to Europe, fully laden from the West Indies. The *Cherubim* proved a success as a privateer, since the High Court of Admiralty papers show that she brought home £2000 worth unspecified 'treasure', as well as sugar, hides, ginger, sarsaparilla, brazil wood and other goods (Andrews, 1964, 243). The vessel is listed as back, working with the Levant Company in 1600 (Table 5.4).

The master of the *Cherabin* was one Thomas Smith and in 1603 he brought a case to court following that ship's loss at sea in a storm. According to his argument, Trinity House had failed to maintain the correct number of markers; where there should have been two, delineating the deeper channel, there was only one. His detailed account states that he

> ....departed downe the River of Thames with the said ship and her ladinge aboute Michelmas last and passed safely through the ffishers Depes and over the Flattes until he came nere a sand called the Last, and perceavinge that there was but only one buoy there, and that the other buoy and beacon which ought to be there were wantinge, he, the said Thomas Smith, fearinge danger for that his ship was depe laden, cast two ankers there, further with his boate and sounded where the water was deepest determininge to bring the ship into the depe water againe & so sett saile & be gon.
>
> And as he and his company were thus labouring, there arose a storme and by the force thereof the ankers came home, and the foresaile being sett was blowen out of the boltrops & the rudder was strucken away and so the ship was cast away with her whole ladinge by reason of the said storme and the wante of the said beacon and buoy ffor as he sayth if the said beacon & buoy had byn there as they ought to have byn, to directe where the depe water was, the ship had passed that danger before the storme had come (Harris, 1969, 169–70).

Thus, for want of a beacon, a fully laden merchantman was lost, right at the start of her outward bound voyage, before she had made the open sea. A sudden violent storm broke off her rudder and ripped off her sails, causing her to drift helplessly until she sank. She was by no means the first or indeed the last ship to suffer such a fate just a day's sailing from London. She shared precisely the same end as the Gresham Ship: she too was a heavily laden trader armed with sakers, on an outward bound voyage. She too seems to have lost her rudder before she sank, since no trace of such was found in the immediate vicinity.

As for the *Cherabin*, there were at least four late 16th-century ships that shared such a name (or a variation thereof) built between 1586 and 1596, and all between 240 and 320 tons (Dietz 1991, 15–16). What can be said with certainty is that the *Cherabin* journeyed great distances before her adventurous life was wrecked by a missing buoy and a rising storm. As such, she epitomized the varied biography of armed merchantmen like the Gresham Ship, as well as the perils they daily faced.

Chapter 6: The Final Voyage

by Mark Beattie-Edwards with Clare Hunt

The Nautical Archaeology Society steps in

The raising of the *Mary Rose* from the depths of the Solent and its transportation to a new home on dry land in Portsmouth in 1982 was one of the most memorable achievements for maritime archaeology in this country. Although not on anything like the same scale, the story of the journey of the remains of the Gresham Ship from Gravesend to a lake on Horsea Island and subsequently to a flooded Leicestershire quarry provides an imaginative alternative. This enterprise could serve as a model for other threatened shipwreck projects with rather more modest resources than the Mary Rose Trust, who raised £2.8m (1982 prices) to record and recover their valuable cargo.

Early in 2004 (three years before the five-year Gresham Ship Project was set up) the Nautical Archaeology Society (NAS) was approached by Sophia Exelby, then Receiver of Wreck at the Maritime and Coastguard Agency (MCA). Following the quite unexpected discovery of the remains of the an Elizabethan merchantman during the Port of London's channel clearance programme, substantial sections of the hull and many artefacts had been recovered from the wreck site, and were currently off loaded at Denton Wharf in Gravesend (Volume I of this monograph, 4, 8). While some of the smaller artefacts would be welcome in museum collections, the Receiver of Wreck asked if the Society would be able to assist in finding a home and a use for the five very much larger composite hull sections, the many single loose timbers, the anchor and the extensive collection of iron bars that constituted part of the merchantman's cargo.

In response the NAS proposed that the timbers and any other material that was not wanted by museums could be placed in a salt water lake in Portsmouth owned and managed by the Ministry of Defence. At the time the NAS was using the Horsea Island Lake on a regular basis as a training facility. If this suggestion were adopted, the timbers would still be accessible for further research and would also be viewed by the diving public.

In a letter dated 8 March 2005 the Receiver of Wreck stated that 'the deposition of these timbers in Horsea Lake is an excellent solution to a difficult problem'. She went on to comment that the timbers from the wreck site were in such a curatorial and legislative no man's land, since:

> there seems to be a gap in the current management of the historic environment. The issue of funding, and continuity of care in particular leaves a gap which cannot presently be bridged. Funding is a key issue as … conservation of material from submerged contexts is inherently very expensive.

She finally went on to say that 'it would be important to for a joint "Princes Channel Wreck Group" to be formed … in order to jointly benefit from steps forward in knowledge and understanding'. A research group was indeed established (this monograph being the fruit of its deliberations) and the timbers from the wreck site were prepared for their next voyage.

Fort Cumberland and Horsea Island 2004–2012 (Plate 25: Figs 6.1; 6.2)

With the approval of the Receiver of Wreck and with the support of the Royal Engineers, the NAS managed to move the timber material, the anchor and some of the iron bars from Denton Wharf in Gravesend, Kent. Acquiring the much needed services of the Royal Engineers was arranged by Marty Hales and Roger Crook (both NAS members) and the move itself was coordinated by Tim Hillsden and the Plant Troop, 24 Training Support Squadron, Royal School of Military Engineering, Chatham. All the material was temporarily stored at the NAS compound in Fort Cumberland until the early August 2004, before being placed in Horsea Island Lake in the first week of September 2004.

Whilst at Horsea the remains of the Gresham Ship were utilized as a recreational diver attraction allowing visitors to see at first hand a 16th-century hull, as well as for NAS training courses on topics such as dendrochronology, photography and videography. However all was not well at Horsea Island and in 2007 the NAS, along with all recreational divers lost access to the water and to the entire site. This was because the MoD felt the security risks of public access – to what is after all a naval base – were too great. Therefore a new home would be needed, if the Gresham Ship was to remain accessible to the diving public.

The Final Voyage 2012

A series of options were considered, but it was not until November 2010 that a viable new home was found in a fresh water lake in Leicestershire. Following an approach by the NAS, Martin Woodward from the National Dive Centre at Stoney Cove sent the following reply by email on 19 November 2010: 'I am delighted to offer the Gresham Ship a new home at Stoney Cove and I will do everything I can to assist the Nautical Archaeology Society'. On the 1 December 2010, Martin travelled to London to attend a meeting of the Gresham Ship Project Steering Group and confirmed Stoney Cove's willingness to provide a home to the material. He agreed to take not only the five wooden hull sections but also the anchor and the iron bar cargo. The reconstituted assemblage would thus once again form the basis of an underwater museum attraction as well as a training site for NAS courses on underwater archaeology. With between 30,000 and 35,000 visitors per year the Gresham Ship Project Steering Group believed this new home would provide a great opportunity for people to experience their heritage in a safe underwater environment.

Operation Gresham
(Plates 25 and 26: Figs 6.3; 6.4; 6.5; 6; 6.7; 6.8)

The transportation of the hull timbers was code-named 'Operation Gresham'. After nearly 17 months of preparation, in May 2012 and with funding support from the PLA and from the Mercers' Company, the NAS arranged the transfer of the remains of the Gresham Ship, a move which would see them transported from Portsmouth, 150 miles north to their new home in Leicestershire. Many preparation dives were carried out in the weeks prior to the lift out of the lake, principally by members of the Portsmouth East branch of the Royal Navy Sub-Aqua Club (PERNSAC). The final lift took place between the 28 May and the 1 June 2012. Once again the hull sections were moved by a team of NAS staff and volunteer divers from the defence contractor QinetiQ (including NAS members) with the assistance of the Royal Engineers. On the Tuesday 29 May 2012 the Gresham ship broke the surface and was loaded onto Royal Engineer transporters for the journey to Leicestershire.

Media interest came from *BBC South Today* and ITV's *Central Tonight* in Leicestershire who both ran stories on the move in their evening news bulletins. The story was covered in the London edition of the *Metro* newspaper and appeared on the Simon Mayo *Drivetime* show on BBC Radio 2 as a Three Word Tuesday – 'Elizabethan Shipwreck Shifted'. Listeners to local BBC radio where even warned that morning to stay off the motorway network just in case the convoy caused disruption to their commute.

Moving the timbers into the water at Stoney Cove on Wednesday 30 May 2012 involved craning them off the Royal Engineer low-loaders and depositing them back underwater on to a 6m deep ledge next to the crane. Because the required crane reach was considerably less than at Horsea Lake it meant that a smaller crane could be used. This meant that it was not necessary to hire a crane, since it was possible to use Stoney Cove Dive Centre's own crane; however, this was required for two days rather than the planned single day.

To move the remains from the ledge to their final resting place they were lifted again, this time using three air filled 250kg marine salvage lifting bags, before being towed by boat to their new permanent home. This might sound like a simple task, but when the heaviest section weighs over 4 tonnes and is an unusual shape as well as being fragile, the difficulty of the exercise cannot be overstated. This is especially true when the dive team were challenged to place them back in the same spatial orientation that they were recorded by the Wessex Archaeology team diving in the Thames in 2003.

One of the composite hull sections (Section 1) only just survived the move and had taken apart in the car park at Stoney Cove before being transported 250m across the lake on a wooden pallet. It then had to be put back together underwater by the divers like a giant jigsaw puzzle – thankfully the visibility in Stoney was so much better than at Horsea Island or this would have proved impossible.

A ten minute film of the move to Stoney Cove, made by NAS member Mark Callaghan of Nautilus Images, was premiered at the NAS Annual Conference in November 2012 and can be seen on the NAS YouTube channel alongside two short clips of local news coverage of the operation. Visit www.youtube.com/user/nautarchsoc

The project team included Mark Beattie-Edwards (NAS), Sara Hasan (Diving Team Coordinator), Alby Hall (Diving Operation Supervisor), Mark Callaghan and Sarah Jessop (from *Nautilus Images* – Video), Martin Davis (Photography) and the principal divers Shaun Samways, Mike Parkin, Mark Hobbs (from *QinetiQ*) and Steve Woodley. Other volunteer divers assisted throughout the project both at Portsmouth and at Stoney Cove. The whole operation took ten days. Later analysis of the diving logbooks showed that 21 divers spent 6496 minutes underwater – that's an astonishing 4½ days.

Costing Operation Gresham

A calculation of the potential costs incurred to transfer the Gresham Ship hull fragments to Stoney Cove suggested that a figure in the region of over £40,000 would be required.

Project management:	£2,500
Project dive team of 10:	£15,000
Crane Hire:	£9,000
Transportation:	£3,000
Equipment hire:	£1,000
Foam and wrappings:	£1,000
Accommodation:	£5,000
Subsistence:	£3,000
Ancillary equipment:	£500
Total:	**£40,000**

In the event the majority of the costs required for Operation Gresham were met by generous contributions in cash and kind from sponsors and from the enthusiastic input of a remarkable team of volunteers. The NAS most gratefully acknowledges the support of the Mercers' Company, University College London, University of Southern Denmark, the Stoney Cove Dive Centre (crane hire, staff and dive team accommodation) and QinetiQ (lifting bags and strops). Especial thanks for help with the transportation are also due to the

Royal Navy and the Field Squadron (Air Support) of 39 Engineer Regiment (based at Waterbeach Barracks in Cambridge), without whom the last voyage of the Gresham Ship could not have been launched.

Operation Gresham has demonstrated is that it does not have to cost the earth to relocate a vessel threatened by necessary redevelopment to a new location where the public can visit and enjoy it. We hope that similar projects elsewhere might adopt or develop such an approach where it represents an appropriate option for the preservation of our nautical heritage.

An Underwater Classroom
(Plate 26: Figs 6.9; 6.10)

Some NAS training course participants may remember undertaking survey exercises recording small objects placed in a swimming pool or perhaps even on a 'little blue boat' with some artefacts in it. But just imagine being able to do such essential training on genuine 16th-century hull remains or on an Elizabethan anchor. This is precisely the use made by the NAS of the Gresham Ship for two inaugural training courses in underwater archaeology at Stoney Cove between the 26 and 28 April 2013. The course participants used the remains as their 'course site' and were challenged to undertake two dimensional and then three dimensional surveys in relation to fixed survey control points. Following the success of this initiative, the new underwater classroom will be the focus for annual training sessions.

In 2013 Martin Woodward from Stoney Cove said:

> 'Stoney Cove first attracted scuba diving pioneers during the 1950s and it's been recognized as The National Diving Centre by the British Sub Aqua Club since 1963. Continued investment in the centre, along with the high quality – and superb features – of its underwater environment mean that Stoney Cove remains an essential entry in any diver's logbook. In 2012, the Gresham Ship's arrival at Stoney Cove created ripples of excitement throughout the diving world and now the UK's first underwater museum exhibit gives over 30,000 visiting divers a year access to a genuine Elizabethan shipwreck.
>
> One of the most popular activities at Stoney Cove has always been the training of underwater archaeologists. Now the Gresham Ship has raised the game. This unique historical artefact is making a considerable contribution to the awareness of our underwater heritage. We are proud to have her'.

The work is still not finished and more visits are being made in 2013 to move the rest of the 49 iron bars to the site and to undertake a photomosaic for an information panel to be located on the quayside at Stoney Cove. Hopefully the 35,000 divers that visit the centre annually will respect the site and not move or damage the material. All the feedback from divers who have already seen the Gresham Ship in its new home has been very positive: many were genuinely amazed that they were able to see at such close quarters the actual remains of an Elizabethan merchantman that had lain in the Thames Estuary for 400 years.

Return to the Thames
With Clare Hunt

But what of the rest of the artefacts? Although the bar iron and the anchor from the Gresham Ship will be displayed underwater at Stoney Cove and the guns will remain with Royal Armouries continuing their desalination treatment, a new home has now been found on dry land for the rest of the finds recovered from the wreck site. Following negotiations in 2011–12 by the Gresham Ship Project Steering Group with Ken Crowe and Clare Hunt of the Southend Museums Service, the artefacts have now been placed in their care. The assemblage will therefore rest alongside other material retrieved by Wessex Archaeology during their monitoring of the navigation channel clearance work conducted by the Port of London Authority in 2003–4 (Ingason *et al.*, 2010; Firth, 2011). These operations included investigations of the wrecks of the *London* (1657–65), the *Dovenby* (1891–1914), the *Atherton* (1918–21), as well as an important group lost during World War II, the SS *Letchworth*, the *Argus*, HMT *Amethyst*, HMS *Ash* and the East Oaze Light Vessel (Ingason *et al.*, 2010; Firth, 2011). Taken together, the stories of these ships as well as our Elizabethan merchantman will provide a remarkable material history of the estuary and its shipping.

Plans are now underway to build a new seafront museum in Southend, actually set into the cliffs, and thus with extensive views over the Thames where the Gresham Ship was lost some 400 years ago. This will be a state-of-the-art complex, with world-class temporary exhibition galleries, a digital theatre, a restaurant and café, as well as shops and exhibition spaces. The museum will bring together all Southend's key collections under one roof, including the nationally important finds from the elite Saxon grave at Prittlewell and the prestigious Beecroft Art Gallery, with its evocative paintings of the Thames. Naturally, maritime archaeology will also be showcased in the new galleries. Indeed, some of the finds from the Gresham ship have already been displayed in the present museum premises, as part of an exhibition that ran over the summer in 2012. The full list of finds from the ship that are now curated by the Southend Museums Service is shown in the Appendix.

Fig. 6.1 Hull timbers at Denton Wharf, 2004 (Nautical Archaeology Society)

Fig. 6.2 The move to Horsea Island, Sept 2004 (Nautical Archaeology Society)

Fig. 6.3 Crane at Horsea Island, as hull timbers are moved again (Nautical Archaeology Society)

Fig. 6.4 Hull timbers laid on to the low loaders at Horsea Island (Nautical Archaeology Society)

Fig. 6.5 (right) Gresham hull Section 4 being lifted from Horsea Lake (Nautical Archaeology Society)

Fig. 6.6 Moving the hull timbers: Stoney Cove 30 May 2012 (© Martin Davies)

Fig. 6.8 Timbers and divers being towed across Stoney Cove (Nautical Archaeology Society)

Fig. 6.9 Underwater classroom: the NAS at Stoney Cove, June2012 (© Martin Davies)

Fig. 6.7 Lifting the hull timbers using air-filled lifting bags (© Martin Davies)

Fig. 6.10 Archaeology student recording the Gresham Ship (© Martin Davies)

Chapter 7: Postscript

The Gresham Ship Project was a university-led research programme that developed under rather unusual circumstances, commencing some five years after the chance discovery of the remains of an Elizabethan merchantman during routine channel maintenance work being conducted by the Port of London Authority. It was thanks to the PLA's responsible approach to the discovery that archaeologists were alerted to the find and subsequently a major research project was initiated. Graduate students and university departments worked together with hands-on studies, using the Gresham Ship materials as a remarkable database for their investigations and dissertations. This was a collaboration between two academic institutions, University College London and the University of Southern Denmark in Esbjerg. As a result of those studies, we now have a detailed assessment of the lost ship and the valuable cargo that sank with her. Rather than focusing on warships, an all too common theme in maritime archaeology, here was a chance to look at merchantmen, the nautical workhorses of the Elizabethan economy. The resulting work has illuminated English construction techniques in an age of transition, continental influence and innovation, considering the process of 'furring' for the first time from an archaeological perspective. The opportunity was also taken to evaluate and develop recording systems, modelling techniques and new approaches to the conservation of maritime objects, particularly the amorphous concretions that feature regularly on wreck sites but are often subsequently neglected. The metallurgical analysis of the metal cargo also developed new directions in, for example, the provenancing of the bar iron as well as in discussion of the trade it represented. A third academic institution also supported the project, for Gresham College in London organized lectures and a conference to extend the archaeological programme. A new home has been found for the significant assemblage from the wreck site with the Southend Museums Service, while the subsequent relocation of the hull timbers at Stoney Cove National Dive Centre provides another major initiative. A new museum exhibit and classroom have been laid out there underwater, thanks to the hard work of the Nautical Archaeology Society, the first such facility in the UK. The vessel remains will thus be in a safe, shallow-water inland location, accessible to many, many more members of the diving public than are able to view the nine protected sites of 16th-century vessels off our coast (Fenwick and Gale, 1998, 146).

The project has also illuminated the central role such ships played in shaping England's history in the 16th century. This is reflected, for example, in the discovery of a remarkable collection of jewellery unearthed during demolition work in Cheapside, over a century ago. This has been dated to the late 16th to early 17th century and throws light on the direct and indirect contact London had with the global traffic and trade networks in the period when the Gresham Ship was operative. Study of the geography reflected in the hoard's precious and semi-precious gemstones is most informative. Serpentine, toadstones and some of the pearls are all from the British Isles, but the rock crystal, amethysts, amber, opals, agates, cornelian, sardonyx, bloodstones and onyx came from all corners of continental Europe. Evidence of the wider world is provided by the malachite and azurite from Russia as well as the lapis lazuli and turquoise, derived from Iran and the Sinai Peninsula. The peridots came from the Red Sea island of Zebriget or St John's Island, the emeralds came across the Atlantic from Colombia, as did the Brazilian topaz and the amazon stone. From the other side of the world, from Sri Lanka, there were chrysoberyl cat's eyes, spinels, moonstones and iolite, while the rubies, chalcedony, diamonds and garnets came from mainland India (Forsyth, 2003, 39).

Such a distribution reflects the early development of trading organizations such as the Muscovy Company (1555), Eastland Company (1579), Levant Company (1581), Barbary Company (1585), East India Company (1600) and the Virginia Company (1606). The Cheapside Hoard therefore represents a global network of merchants, middlemen and international entrepots with which Elizabethan ships lawfully engaged. It may also reflect the less reputable means by which such treasures were amassed, as the study of a ceramic assemblage from a possible 'pirate' community in Limehouse reminds us (Killock *et al.*, 2005). What percentage of the Cheapside Hoard reflects the profits of privateering is impossible to calculate, but will be significant. The catalogue compiled by the High Court of Admiralty of cargoes captured by the belligerent crews of Elizabethan merchantmen in the 1590s includes bullion and pearls by the *Harry & John*, unspecified 'treasure' by the *Cherubim* and by the *Moonshine*, silver, gold and jewels by the *Julian*, as well as precious stones by the *Bark Randall*, for example (Andrews, 1964, 243–7). Then, in 1592, there was the capture by the Earl of Cumberland's fleet (bankrolled by the Queen and the City of London) of the great Portuguese carrack, the *Madre de Dios*, stuffed with jewels, silk and other precious goods (Andrews, 1964, 73).

Clearly the very considerable wealth of the Protestant port of London (and indeed the rest of the island nation) relied crucially upon its shipping. This was a time when armed merchantmen like the Gresham Ship were opening up new trade routes, exploring new worlds, circumnavigating the globe, serving as plundering privateers or taking on the Spanish Armada. For good or for ill, such ships and their redoubtable crews were the making of Elizabethan England.

Appendix: Finds List

Object no. (GSP)	Layer no.	Material	Type of object	Dimensions (mm)	Object description	Conservator & Cons lab no.	Treatment summary	Archive location
10	105	IRON	BAR	Unknown	n/a	n/a	Retained with hull	Stoney Cove
11	100	WOOD	BARREL STAVE	424 x 94 x 25	Complete barrel stave with complete u-shaped lid groove across width of wider end. Flat length worked from tangential section and cross-section corresponds to transverse plane. Dark brown in colour. Species: *Quercus*	Kelly Domoney	Desalinated, PEG immersion, freeze drying, possible gap filling after drying	Southend Museums
12	100	CONCRETION	CONCRETION	220 x 190 x 50	Iron sulphide concretion with outer $CaCO_3$ crust	Natsumi Henzan 8548	Completely processed, no finds	Rubble fragments discarded
13	100	CERAMIC	BRICK BUILDING MATERIAL	60 x 60 x 45		Libby McCormick	Desalinated and dried	Southend Museums
14	100	LEAD ALLOY	SALT HOLDER	74 x 65 x 64	Low-lead pewter with moulded decoration on rim, centre and base. One side of upper section bent upwards. Alloying elements identified by XRF include tin, copper, antimony, iron and lead	Kelly Domoney	Desalinated and dried	Southend Museums
15	101	LEAD	SHEET	200 x 95 x 5	Lead sheathing with nail holes and nail head impressions around edge; 10 nail holes, square with diameter 1–2 mm	Libby McCormick	Desalinated and dried	Southend Museums
16	124	LEAD	INGOT	530 x 210 x 140	Oval lead ingot with three identical stamp marks, initials 'EB'	Kelly Domoney	Desalinated and dried	Southend Museums
17	104	IRON	BAR	1690 x 350 x 350	n/a	n/a	n/a	Stoney Cove
18	104	IRON	BAR	1970 x 80 x 30	n/a	n/a	n/a	Stoney Cove
19	101	IRON	BAR	Unknown	n/a	n/a	Retained with hull	Stoney Cove

Object no. (GSP)	Layer no.	Material	Type of object	Dimensions (mm)	Object description	Conservator & Cons lab no.	Treatment summary	Archive location
20	100	CONCRETION	CONCRETION	700 x 390 x 340	Heavy concretion with solid $CaCO_3$ outer crust interspersed with marine sediment and shells	Hazel Gardiner, Libby McCormick	Two cavities cast (GSP20.5 and GSP20.6)	Rubble fragments discarded
20.1		SILVER	SPOON	152 x 83 x 14	Silver spoon with ceramic fragment attached	Libby McCormick	Cleaned, desalinated and dried	Southend Museums
20.2		CERAMIC	FRAGMENTS			Libby McCormick	Cleaned, desalinated and dried	Southend Museums
20.3		BONE	FRAGMENT			Libby McCormick	Cleaned, desalinated and dried	Southend Museums
20.4		IRON	FRAGMENTS		Group of small iron fragments	Libby McCormick	Cleaned, desalinated and dried	Southend Museums
20.5		POLYESTER RESIN	CAST		Polyester resin cast containing Iron weighting powder	Libby McCormick	Cast and dried	Southend Museums
20.6		WOOD/ POLYESTER RESIN	CAST			Libby McCormick	Cast contains wood; desalinated and dried	Southend Museums
21	100	STONE	BALLAST STONE	250 x 230 x 190		Kelly Domoney	Dried	Southend Museums
22	104	IRON	BAR	1970 x 100 x 100		n/a	n/a	Stoney Cove
23	104	IRON	BAR	Length: 1830		n/a	Retained with hull	Stoney Cove
24	104	IRON	BAR	Length: 1870		n/a	Retained with hull	Stoney Cove
25	100	IRON	CONCRETION	220 x 130 x 80	Iron sulphide concretion with outer $CaCO_3$ crust	Sylvia Haliman 8550	Complete; cast taken	Rubble discarded
25.1		POLYESTER RESIN	NAIL CAST		Nail or deck plank spike showing change of form from square cross-section to round cross-section. Bottom tip appears to be broken, could have been fastened to one side and hooked nail	Sylvia Haliman 8550	Cast and reconstructed	Southend Museums
26	100	IRON	CONCRETION	310 x 50 x 50	Concretion containing modern steel cable fragment	Lyda Muros 8541	Part cleaned, Desalinated, dried	Southend Museums

Object no. (GSP)	Layer no.	Material	Type of object	Dimensions (mm)	Object description	Conservator & Cons lab no.	Treatment summary	Archive location
27	100	IRON	CONCRETION	410 x 150 x 160	Heavy iron sulphide concretion with cavity of corroded wrought iron chain links	Kelly Domoney	Casts taken, GSP27.1–27.3, concretion section dried GSP27.4–27.5.	Rubble discarded
27.1		POLYESTER RESIN	CAST	Length: 96 (uncurved length: 193) diameter: 15	Casts of chain end cavities, made from polyester resin with iron filler	Kelly Domoney	Cast and dried	Southend Museums
27.2		EPOXY	CAST	Length: 139 diameter: 15	Casts of chain end cavities, made from epoxy resin with pigments	Kelly Domoney	Cast and dried	Southend Museums
27.3		EPOXY	CAST	Length: 82 diameter: 15	Casts of chain end cavities made from epoxy resin with pigments	Kelly Domoney	Cast and dried	Southend Museums
27.4		CONCRETION	CONCRETION		Central section of concretion	Kelly Domoney	Desalinated and dried	Southend Museums
27.5		CONCRETION	CONCRETION		Section of concretion	Kelly Domoney	Desalinated and dried	Southend Museums
28	150	IRON	CANNON	Length: 1980, cascable length: 220, depth base ring: 300, depth muzzle face: 160, bore: 760	Inscribed with initials T G, calibre mark/weight 8-0-0, with grasshopper motif. Made in Thomas Gresham's foundry in Mayfield, Sussex between 1567 and 1579 (designated PC1 in Volume I)	n/a	Undergoing desalination	Fort Nelson R Armouries
29	100	CONCRETION	CONCRETION	103 x 46 x 34			Undergoing processing (2013)	
30	104	IRON	BAR	1030 x 50 x 40		n/a	n/a	Stoney Cove
31	126	POTTERY	FRAGMENT	69 x 52 x 11	Low fired non-glazed red ceramic. Ceramic manufactured from a grey clay. Red outer surface indicates firing in oxidizing conditions. Quartz inclusions 0.1–2 mm in size.	Kelly Domoney	Desalinated and dried	Southend Museums

Object no. (GSP)	Layer no.	Material	Type of object	Dimensions (mm)	Object description	Conservator & Cons lab no.	Treatment summary	Archive location
32	126	WOOD	FRAGMENT	91 x 27 x 20	Worked piece of oak with corroded iron peg through width and iron concretion on surface (*Quercus*) Underside has carved notch across the width. Wood is complete in shape, rectangular in plan view and rhomboid in side profile. Cross-section worked from transverse plane of the tree	Kelly Domoney	Desalinated, PEG and TEA immersion, freeze dried	Southend Museums
33	101	TIN	INGOT	490 x 30 x 15	Grey coloured and bar-shaped	Kelly Domoney	Desalinated and dried	Southend Museums
34	119	POTTERY	FRAGMENT	109 x 74 x 7	Pottery fragment of high-fired ceramic with grey inclusions and red-coloured glaze	Kelly Domoney	Cleaned, desalinated and dried	Southend Museums
35	126	LEATHER	SHOE	195 x 65 x 4	Waterlogged vegetable tanned leather shoe sole. Stitch marks can be seen on perimeter of sole	Cymbeline Storey 8432	Cleaned, desalinated, glycerol impregnation, freeze-dried; reconstructed	Southend Museums
36	126	COPPER ALLOY	FRAGMENT	60 x 30 x 10	Copper alloy fragment		Desalinated and dried	Southend Museums
37	104	IRON	BAR	Unknown		n/a	Retained with hull	Stoney Cove
38	104	IRON	BAR	820 x 90 x 70		n/a	n/a	Stoney Cove
39	104	IRON	BAR	106 0 x 120 x 50		n/a	n/a	Stoney Cove
40	104	IRON	BAR	2000 x 160 x 100		n/a	n/a	Stoney Cove
41	104	IRON	BAR	Unknown		n/a	Retained with hull	Stoney Cove
42	104	IRON	BAR	1960 x 100 x 100		n/a	n/a	Stoney Cove
43	104	IRON	BAR	1950 x 100 x 100		n/a	n/a	Stoney Cove

Object no. (GSP)	Layer no.	Material	Type of object	Dimensions (mm)	Object description	Conservator & Cons lab no.	Treatment summary	Archive location
44	104	IRON	BAR	196 x 100 x100		n/a	n/a	Stoney Cove
45	104	IRON	BAR	2000 x 250 x 250		n/a	n/a	Stoney Cove
46	104	IRON	BAR	1900 x 160 x 50		n/a	n/a	Stoney Cove
47	104	IRON	BAR	Unknown		n/a	n/a	Stoney Cove
48	104	IRON	BAR	1140 x 400 x 400		n/a	n/a	Stoney Cove
49	104	IRON	BAR	Unknown		n/a	Retained with hull	Stoney Cove
50	101	IRON	BAR	Unknown		n/a	Retained with hull	Stoney Cove
52	105	IRON	BAR	Unknown		n/a	Retained with hull	Stoney Cove
53	101	IRON	BAR	Unknown		n/a	Retained with hull	Stoney Cove
54	101	IRON	BAR	Unknown		n/a	Retained with hull	Stoney Cove
55	126	LEATHER	SHOE	138 x 210 x 60	Waterlogged vegetable tanned bovine leather. Fragmentary heel and sole of shoe with calcium carbonate	Luisa Duarte 8451	Desalinated, cleaned, impregnated with glycerol/TEA and freeze-dried	Southend Museums
56	104	IRON	BAR	Unknown	n/a	n/a	Retained with hull	Stoney Cove
57	104	IRON	BAR	Unknown	n/a	n/a	Retained with hull	Stoney Cove
58	106	GLASS	FRAGMENT	58 x 87 x 5	Fragment of a free blown cylindrical straight-sided brown/blue glass bottle with pushed in dome-shaped indented base	Kelly Domoney	Desalinated and dried	Southend Museums
59	104	IRON	BAR	Unknown	n/a	n/a	Retained with hull	Stoney Cove
60	124	LEAD	FRAGMENT	70 x 10 x 10		Libby McCormick	Desalinated and dried	Southend Museums
61	104	IRON	BAR	Unknown	n/a	n/a	Retained with hull	Stoney Cove
62	104	IRON	BAR	Unknown	n/a	n/a	Retained with hull	Stoney Cove
63	104	IRON	BAR	Unknown	n/a	n/a	Retained with hull	Stoney Cove

Object no. (GSP)	Layer no.	Material	Type of object	Dimensions (mm)	Object description	Conservator & Cons lab no.	Treatment summary	Archive location
64	104	IRON	BAR	Unknown	n/a	n/a	Retained with hull	Stoney Cove
65	104	IRON	BAR	Unknown	n/a	n/a	Retained with hull	Stoney Cove
66	104	WOOD/ CONCRETION	CONCRETION	280 x 90 x 140	Iron sulphide concretion with outer $CaCO_3$ crust. Wood fragment and resin cast. Cast dimensions: 24 x 50	Sarah Nettleton 8544	Consists of concretion. Wood and polyester resin casts. Desalinated, PEG/TEA impregnation and freeze dried	Southend Museums
68	101	CONCRETION	CONCRETION	270 x 240 x 120	Iron sulphide concretion with outer $CaCO_3$ crust	Ashley Lingle 8542	Processed and dry, no finds	Rubble discarded
69	125	IRON	BAR	Unknown	n/a	n/a	Retained with hull timbers	Stoney Cove
70	124	CONCRETION	CONCRETION	560 x 270 x 170	Iron sulphide concretion with outer $CaCO_3$ crust	Alex Solomon 8539	Processed and dried, no finds	Rubble discarded
71	100	STONE	BALLAST STONE	480 x 330 x 160	n/a	Kelly Domoney n/a	Desalinated and dried	Southend Museums
72	100	STONE	BALLAST STONE	480 x 260 x 160	n/a	Kelly Domoney n/a	Desalinated and dried	Southend Museums
73	104	IRON	BAR	Unknown	n/a	n/a	Retained with hull	Stoney Cove
74	104	IRON	BAR	Unknown	n/a	n/a	Retained with hull	Stoney Cove
75	104	IRON	BAR	Unknown	n/a	n/a	Retained with hull	Stoney Cove
76	104	IRON	BAR	850 x 120 x 100	n/a	n/a	n/a	Stoney Cove
77	104	IRON	BAR	Unknown	n/a	n/a	Retained with hull	Stoney Cove
78	104	IRON	BAR	1950 x 100 x 100	n/a	n/a	n/a	Stoney Cove
79	104	IRON	BAR	Unknown	n/a	n/a	Retained with hull	Stoney Cove
80	101	LEATHER	BOOT FRAGMENT	630 x 500 x 5	Four pieces of vegetable tanned calf leather; one large and three smaller fragments. Initially thought to represent remains of a jerkin, now identified as upper of an over-the-knee boot	Katy Smith 8427	Desalinated, cleaned, impregnated with glycerol and freeze-dried	Southend Museums

Object no. (GSP)	Layer no.	Material	Type of object	Dimensions (mm)	Object description	Conservator & Cons lab no.	Treatment summary	Archive location
81	101	TIN	INGOT	600 x 30 x 10	Grey coloured and bar-shaped	Kelly Domoney	Desalinated and dried	Southend Museums
82	111	WOOD	FRAGMENT	165 x 39 x 19	Small sawn wood fragment worked on four sides, with wedge-like elongated pyramidal shape. Cross-section corresponds with tree's transverse plane. Species *Quercus*.	Kelly Domoney	Desalinated, PEG immersion and freeze dried.	Southend Museums
83	100	CONCRETION	CONCRETION	70 x 60 x 50	Iron sulphide concretion containing impression of herring bone pattern	Rosemary Jeffreys 8549	Desalinated, dried and cast taken	Southend Museums
83.1	100	SILICONE	CAST		Silicone rubber impression taken		Cast and dried	Southend Museums
84	103	LEAD	INGOT	590 x 190 x 95	Oval lead ingot with three identical stamp marks, initials 'EB'	Kelly Domoney	Desalinated and dried	Southend Museums
85	103	LEAD	INGOT	600 x 200 x 90	Oval lead ingot with three identical stamp marks, initials 'EB'	Kelly Domoney	Desalinated and dried	Southend Museums
86	103	CONCRETION	CONCRETION	310 x 200 x 140	Iron sulphide concretion with outer $CaCO_3$ crust	Kelly Domoney	Complete; processed, section of concretion retained, showing imprint of iron bar GSP86.2 (dry), casts GSP86.2 and GSP86.3 taken	Rubble discarded.
86.1	103	TIN ALLOY	SALT HOLDER	108 x 77 x 33	Low-lead pewter salt holder with incised linear design. Alloying elements identified by XRF include tin, antimony, iron, copper and lead.	Kelly Domoney	Cleaned, desalinated and dried	Southend Museums
86.2	103	SILICONE	CAST		Silicone rubber cast of wrought iron bar cavity	Kelly Domoney	Cast of bar cavity created	Southend Museums
86.3	103	SILICONE	CAST		Silicone rubber cast of wrought iron nail cavity	Kelly Domoney	Cast of nail cavity created	Southend Museums
86.4	103	SILICONE	CAST		Silicone rubber cast of wrought iron nail cavity	Kelly Domoney	Cast of nail cavity created	Southend Museums

Object no. (GSP)	Layer no.	Material	Type of object	Dimensions (mm)	Object description	Conservator & Cons lab no.	Treatment summary	Archive location
86.5		SILICONE	CAST		Silicone rubber cast of wrought iron nail cavity	Kelly Domoney	Cast of nail cavity created	Southend Museums
86.6		WOOD	FRAGMENT		Small fragment of wood located inside base of GSP86.1	Kelly Domoney	Desalinated and dried	Southend Museums
86.7		TIN ALLOY	SPOON	Height: 79, head width: 30, depth: 2, handle: length: 68, width: 4–8, depth: 3–4	Tin alloy spoon with moulded decoration	Kelly Domoney	Cleaned, desalinated and dried	Southend Museums
87	103	CERAMIC BUILDING MATERIAL	BRICK	220 x 100 x 53		Libby McCormick	Desalinated and dried	Southend Museums
88	127	LEATHER	SHOE	460 x 180 x 130	Waterlogged leather shoe; composite comprised of wood, leather and concretion	Robyn Haynie 8428	Desalinated, PEG/TEA immersion, freeze dried, possible gap filling after drying	Southend Museums
89	101	LEAD	PIPE	1900 x 50 x 50		Dean Sully	Air dried	Southend Museums
90	101	CONCRETION	CONCRETION	410 x 300 x 140	Iron sulphide concretion with outer $CaCO_3$ crust.	Kelly Domoney	Chain cavities cast with Polyester resin	Southend Museums
90.1	101	IRON	CHAIN LINKS		Two wrought iron chain links	Kelly Domoney	Cleaned, desalinated and dried	Southend Museums
91	101	TIN	INGOT	360 x 15 x 10	Grey in colour and bar-shaped	Kelly Domoney	Desalinated and dried	Southend Museums
92	127	IRON	BAR	1120 x 80 x 40			Retained with hull	Stoney Cove
93	127	CONCRETION	CONCRETION	600 x 300 x 300	Iron sulphide concretion with outer $CaCO_3$ crust	Julia Tubman 8545	Iron nail excavated, desalinated and dried	Southend Museums
93.1	127	WOOD	FRAGMENT		Eroded wood fragment	Julia Tubman 8545	Desalination impregnation with PEG treatment and freeze-dried	Southend Museums

Object no. (GSP)	Layer no.	Material	Type of object	Dimensions (mm)	Object description	Conservator & Cons lab no.	Treatment summary	Archive location
94	101	CONCRETION	CONCRETION	210 x 210 x 150	Iron sulphide concretion with outer $CaCO_3$ crust	Kelly Domoney	Processed to chain link voids, desalinated and dried	Southend Museums
95	127	IRON/ POLYESTER RESIN	CONCRETION/ CAST	230 x 175 x 120	Remains of iron nail core and void cast with polyester resin and reconstructed (dimensions 160 x 22 x 25).	Daniela Pedroza de Boos 8546	Concretion processed, composite iron concretion and resin cast, desalinated and dried	Southend Museums
96	103	CONCRETION	CONCRETION	180 x 160 x 150	Iron sulphide concretion with outer $CaCO_3$ crust. Remains of iron nail void cast with Polyester resin and reconstructed	Julie Chang 8547	Concretion processed, cast taken	Rubble discarded
96.1	103	POLYESTER RESIN	CAST		Polyester concretion cast of iron nail	Julie Chang 8547		Southend Museums
97	103	CONCRETION	CONCRETION	277 x 106 x 140	Iron sulphide concretion with outer $CaCO_3$ crust.	Lucie Monot 8543	Processed and dry, no finds. Rubble fragments retained	Southend Museums
98	102	WOOD	GUN CARRIAGE	630 x 480 x 580	Part of gun carriage made from elm, with iron fastenings and concreted round object on top, found close to GSP201	Graeme McArthur, Julia Tubman	Documentation and corrosion removal to reveal surface detail	Stoney Cove
99	102	CONCRETION	CONCRETION	390 x 160 x 200	Heavy concretion with solid $CaCO_3$ outer crust interspersed with marine sediment and shells, from area around GSP201. Main fragment revealed partially preserved hook and cavity (GSP99.1), Remaining fragment cleaved into five pieces to reveal metal pin transecting wooden or thin metal plate – latter has impression on surface	Kelly Domoney, Libby McCormick	Casts GSP99.1, GSP99.2 and GSP99.3 taken. 5 fragments, in 3 parts remain	Southend Museums

Object no. (GSP)	Layer no.	Material	Type of object	Dimensions (mm)	Object description	Conservator & Cons lab no.	Treatment summary	Archive location
99.1	102	IRON and POLYESTER RESIN	CAST	Length: 255 (Uncurved length: 415) diameter: 28	Cast of a single-prong iron hook cavity containing original wrought iron core. Cast made from polyester resin with iron filler	Kelly Domoney	Void cast, hook reconstructed and dried	Southend Museums
99.2	102	SILICONE	CAST		Fragment of concretion with incised decoration	Libby McCormick	Mould of incised decoration created	Southend Museums
99.3	102	SILICONE	CAST		Fragment of concretion with incised decoration	Libby McCormick	Mould of incised decoration created	Southend Museums
200	102	CONCRETION	CONCRETION	480 x 190 x 190	Iron sulphide concretion with outer $CaCO_3$ crust, from area around GSP201; medium weight wrought iron chain links preserved as voids	Amanda Watts 8538	Concretion processed, cast GSP200.1 taken. One section of concretion retained; desalinated and dried	Southend Museums
200.1		POLYESTER RESIN	CAST		Cast remains of five iron chain links in Polyester resin	Amanda Watts 8538	Cast and dried	Southend Museums
201	102	IRON	CANNON	Length 2150	Designated PC2 in volume I	n/a	Undergoing desalination	Fort Nelson R Armouries
201.1		COPPER ALLOY/ CONCRETION	BOWL and CONCRETION	Diameter: 4318	Concreted onto canon GSP201, copper-alloy bowl and concretion with impression of canon. XRF surface analysis indicated alloy was brass (58–70 copper (Cu); 25–31% zinc (Zn)	Lisa Barter 8782	Cleaned, desalinated and dried	Southend Museums
201.2		CERAMIC/ CONCRETION	CERAMIC JAR and CONCRETION		Concreted onto canon GSP201, fragment of a ceramic salt glazed jar attached to concretion	Rebecca Pont 8551	Cleaned, desalinated and dried	Southend Museums
201.3	102	FIBRES	TEXTILE	125 x 100 x 3	Concreted onto canon GSP201, fragment of modern synthetic fabric	Lyda Muros 8597	Cleaned, desalinated and dried	Southend Museums
202	102	CONCRETION	CONCRETION	560 x 340 x 330	Extremely large and heavy oval shaped concretion.	Kelly Domoney	Completely processed now GSP202.1–GSP202.3	

Object no. (GSP)	Layer no.	Material	Type of object	Dimensions (mm)	Object description	Conservator & Cons lab no.	Treatment summary	Archive location
202.1	102	IRON	POWDER CHAMBER	450 x 250 x 250 420 x 190 x 180	Cast iron body with central wrought iron strap securing in place wrought iron handles; fuse hole is visible on side of barrel	Kelly Domoney	Cleaned, desalinated, dried	Southend Museums
202.2	102	IRON AND WOOD	IRON HAMMER HEAD WITH PARTIAL WOODEN HANDLE	107 x 120 x 35	Iron hammer head with partial wooden handle	Rafael Lopez Bravo 8833	Desalinated, PEG/TEA impregnation and freeze-dried	Southend Museums
202.3	102	CONCRETION/ FIBRE	CONCRETION/ ROPE		Small concretion with embedded rope fragment	Libby McCormick	Desalinated and dried	Southend Museums
203	102	CONCRETION	CONCRETION	550 x 390	Iron sulphide concretion with outer $CaCO_3$ crust		Complete; processed, object GSP203.1 retrieved/cast	Rubble discarded
203.1	102	IRON and POLYESTER RESIN	IRON BOLT and CAST	420 x 190 x 180	Cast of iron bolt cavity containing original iron core. Cast made from polyester resin with iron filler	Graeme McArthur 8540	Void cast, bolt reconstructed, desalinated and dried	Southend Museums
204	125	WOOD	PIKE	310 x 55 x 27	Pike shaft, octagonal and concreted pike head. Species: *Quercus*. (Associated with GSP205)	Graeme McArthur 8617	Cleaned, PEG impregnation, freeze dried, possible gap filling after drying	Southend Museums
205	125	LEATHER	SHOE	280 x 140 x 60	Sole of leather shoe, concreted to GSP204	Morgan Nau 8434	Cleaned, PEG/TEA impregnation, freeze dried.	Southend Museums
206	125	LEATHER	SHOE	116 x 91 x 25 after treatment: 113 x 87 x 26	Part of leather shoe insole, found underneath GSP205	Libby McCormick 8438	Desalinated, cleaned, impregnated with glycerol and freeze-dried	Southend Museums
207	125	WOOD	FRAGMENT	14 x 35 x 7	Thin, semi-circular fragment of worked wood. Possibly section of barrel lid. Main surface worked from transverse section of tree. Species: pine	Kelly Domoney	Desalinated, PEG immersion, freeze dried, possible gap filling after drying	Southend Museums

Object no. (GSP)	Layer no.	Material	Type of object	Dimensions (mm)	Object description	Conservator & Cons lab no.	Treatment summary	Archive location
208	101	COPPER ALLOY	JUG	110 x 130 x 90	Main alloying elements: copper, zinc, lead, tin, antimony (exposed metal analysed by XRF)	Libby McCormick	Cleaned, desalinated, dried	Southend Museums
209	102	CONCRETION	CONCRETION	250 x 210 x 120	Iron sulphide concretion with outer $CaCO_3$ crust. From area around GSP201. Bulbous shaped concretion with wrought iron bar and cavity. Cavity 50% of concretion volume	Libby McCormick, Francesca Guiducci 8831	Concretion processed, GSP209.1 wrought iron pin with iron core and resin cast. GSP209.2–GSP209.4 are resin casts of small voids. Desalinated and dried	
210	102	IRON	BOLT	200 x 80 x 70	Iron sulphide concretion with outer $CaCO_3$ crust. From area around GSP201.	Libby McCormick, Carmen Vida 8832	Semi-processed. Concretion fragment with bolt retained, desalinated and dried. Voids in other concretion fragment were cast, creating GSP210.1–GSP210.4 polyester resin casts of square sectioned nails	Southend Museums
211	102	FIBRE	ROPE	180 x 20 x 20	Fragment of rope connected to concretions around cannon GSP201 and therefore associated with copper-alloy bowl (GSP201.1) and ceramic jar (GSP201.2). A z-spun, S-ply, Z-cabled (zS2(Z)3) rope	Sylvia Haliman 8609	Cleaned, desalinated and dried	Southend Museums
212	150	IRON	ANCHOR					Stoney Cove
213	150	TIN	INGOT	570 x 20 x 5	2 x metal ingots of uniform dimensions	Kelly Domoney	Desalinated and dried	Southend Museums

Object no. (GSP)	Layer no.	Material	Type of object	Dimensions (mm)	Object description	Conservator & Cons lab no.	Treatment summary	Archive location
215	150	POTTERY	FRAGMENT	160 x 150 x 8	Fragment of low-fired dry coarse buff coloured ceramic with fine sand inclusions: glazed Spanish olive jar. Resin layer on interior surface analysed using FTIR; resulting bond peaks were compared with species of pine-resin	Kelly Domoney	Cleaned, desalinated and dried	Southend Museums
216	150	CONCRETION/ FIBRE	CONCRETION/ ROPE	130 x 110 x 90	Iron sulphide concretion containing rope.	Kelly Domoney	Dried, rope and concretion retained	Southend Museums
217	150	CONCRETION/ FIBRE	CONCRETION/ ROPE	4 fragments. 180 x 120 x 40; 80 x 70 x 30; 90 x 40 x 20; 50 x 20 x 20	Iron sulphide concretion containing rope	Kelly Domoney	Dried, rope and concretion retained	Southend Museums
218	150	CONCRETION/ FIBRE	CONCRETION/ ROPE	160 x 130 x 90	Iron sulphide concretion with outer $CaCO_3$ crust	Kelly Domoney	Dried, rope and concretion retained	Southend Museums
219	150	IRON	CANNON			n/a		Fort Nelson R Armouries
220	150	IRON	CANNON			n/a		Fort Nelson R Armouries
WA 54135		WOOD	TREENAIL	160 x 35 (diameter)	Dark brown treenail; one end fractured. Complete end has v-shaped groove, possible tool mark. Species; *Quercus*	Kelly Domoney	Desalinated, PEG immersion. freeze drying, possible gap filling	Southend Museums
WA Brick		BUILDING MATERIAL	BRICK	100 x 80 x 50		n/a	Desalinated and dried	Southend Museums

Bibliography

Volume I of this two-volume monograph has been published as:

Auer, J., and Maarleveld, T.J., eds, 2014, *The Gresham Ship Project: A 16th-Century Merchantman Wrecked in the Princes Channel, Thames Estuary: Volume I: Excavation and Hull Studies*, BAR British Series 602 (NAS Monograph Series No 4), Oxford.

Adams, J., 2000, 'Ships and boats as archaeological source materials', *World Archaeology*, **32(3)**, 292–310.

Adams, J., 2003, *Ships, Innovations and Change: aspects of carvel shipbuilding in northern Europe 1450–1850*, Stockholm Marine Archaeology Report no 3, Stockholm.

Afonso, S.L., *et al.*, ed., 1998, *Nossa Senhora dos Mártires: the last voyage*, Verbo, Lisbon.

Agnasdottir, A., ed., 2001, *Voyages and Exploration in the North Atlantic from the Middle Ages to the XVIIth Century*, University of Iceland Press, Reykjavik.

Alpern, S.B., 1995, 'What Africans Got for Their Slaves: a master list of European trade goods', *History in Africa*, **22**, 5–43.

Alsop, J., 1996, 'William Towerson's Rutter for the Margate-Emden Navigation, 1564', Mariner's *Mirror*, **82(2)**, 154–8.

Andrews, K., 1964, *Elizabethan Privateering during the Spanish War 1585–1603*, Cambridge University Press, Cambridge.

Archer, I., Barron, C. and Harding, V., eds, 1988, *Hugh Alley's Caveat: the markets of London in 1598*, London Topographical Society, London.

Arnold, J.B. and McAllister, M., 1998, 'Artefact replication by stereolithography', *International Journal of Nautical Archaeology*, **27(2)**, 160–5.

Ashley Smith, J., 2009, 'The Basis of Conservation Ethics', in Richmond, A., and Bracker, A., eds, *Conservation: principles, dilemmas and uncomfortable truths*, Butterworth-Heinemann, Oxford, 6–24.

Auer, J. and Firth, A., 2007, 'The Gresham Ship: an interim report on a 16th-century wreck from the Princes Channel, Thames Estuary', *Journal of Post Medieval Archaeology*, **41(2)**, 222–41.

Avrami, E., Mason, R. and de la Torre, M., 2000, *Values and Heritage Conservation*, The Getty Conservation Institute, Los Angeles, CA.

Avrami, E., 2009, 'Heritage, Values, and Sustainability', in Richmond, A., and Bracker, A., eds, *Conservation: principles, dilemmas and uncomfortable truths*, Butterworth-Heinemann, Oxford, 177–83.

Awty, B.G., 1987, 'The origin of the blast furnace: evidence from Francopohone areas', *Historical Metallurgy*, **21(2)**, 96–9.

Awty, B.G., 2006, 'The elusive Walloon finery forges of Liège', *Historical Metallurgy*, **40(2)**, 129–37.

Awty, B.G., 2008, 'The Austrian lift-hammer – its probable Walloon origin', *Historical Metallurgy*, **42(1)**, 12–22.

Ayres, J. and Wroe-Brown, R., 2002, *The London Millennium Bridge: excavation of the medieval and later waterfronts at Peter's Hill, City of London, and Bankside, Southwark*, Museum of London Archaeology Service Archaeological Studies series 6, London.

Baldwin, R., 2001, 'The testing of a New Academic Trinity for the Northern Passages', in Agnasdottir, A., ed., *Voyages and Exploration in the North Atlantic from the Middle Ages to the XVIIth Century*, University of Iceland Press, Reykjavik, 61–98.

Ballard, M. and Skals, I., 1996, 'Masking and misinterpreting cotton fibres: dangers associated with fibre analysis of archaeological textiles', in Roy, A., and Smith, P., eds, *Archaeological Conservation and Its Consequences, Preprints of the Contributions to the Copenhagen Congress, 26–30 August 1996*, International Institute for Conservation, London, 6–10.

Banbury, P., 1971, *Shipbuilders of the Thames and Medway*, David and Charles, Newton Abbot.

Barker, D., 2003, 'Conservation of metals', in Jones, M., ed., *For Future Generations: conservation of a Tudor maritime collection*, Mary Rose Trust, Portsmouth, 75–94.

Barraclough, K.C., 1984, *Steelmaking before Bessemer, vol. 1, Blister Steel: the birth of an industry*, The Metals Society, London.

Barron, C., Coleman, C. and Gobbi, C., 1983, 'The London Journal of Alessandro Magno, 1562', *London Journal*, **9(2)**, 136–52.

Barry, B., 1998, *Senegambia and the Atlantic Slave Trade*, Cambridge University Press, Cambridge.

Beck, C. and Borromeo, C., 1990, 'Ancient Pine Pitch: technological perspectives from a Hellenistic shipwreck', in Biers, W.R., and McGovern, P.E., eds, *Organic contents of ancient vessels: materials analysis and archaeological investigation*, MASCA Research Papers in Science and Archaeology, **7**, 51–8.

Beier, A.L., and Finlay, R., eds, 1986, *The Making of the Metropolis: London 1500–1700*, Longman, London.

Bindoff, S.T., 1950, *Tudor England*, Penguin, Harmondsworth, Middlesex.

Birch, T. and Martinón-Torres, M., forthcoming. 'The iron bars from the "Gresham Ship": employing multivariate statistics to further slag inclusion analysis of ferrous objects', in Belford P., Bayley, J., and Crossley, D., eds, *Iron and Ironworking: Proceedings of the Archaeometallurgy Conference, Bradford, November 2010*, Historical Metallurgy Society, London.

Björdal, C.G., Nilsson, T., and Petterson, R., 2007, 'Preservation, storage and display of waterlogged wood and wrecks in an aquarium: "Project Aquarius"', *Journal of Archaeological Science*, **34(7)**, 1169–77.

Björkenstam, N., 1982, 'Technology, production, costs', in Pipping, G. ed., *Iron and Steel on the European Market in the 17th Century: a contemporary Swedish account of production forms and marketing*, The Historical Metallurgy Group of the Swedish Ironmasters' Association, Stockholm, 23–51.

Blakelock, E., Martinón-Torres, M., Veldhuijzen, H. A., and Young, T., 2009, 'Slag inclusions in iron objects and the quest for provenance: an experiment and a case study', *Journal of Archaeological Science*, **36(8)**, 1745–57.

Blatherwick, S. and Bluer, R., 2009, *Great houses, moats and mills on the south bank of the Thames: medieval and Tudor Southwark and Rotherhithe*, Museum of London Archaeology Service monograph 47, London.

Blick, C.R., *et al.*, 1991, 'Early blast-furnace notes', *Historical Metallurgy*, **25(1)**, 47–68.

Bluer, D., 1993, 'Excavations at Abbots Lane, Southwark', *London Archaeologist*, **7(3)**, 59–66.

Bluestone, D., 2000, 'Challenges for Heritage Conservation and the Role of Research on Values', in Avrami, E., Mason, R. and de la Torre, M., eds, *Values and Heritage Conservation*, The Getty Conservation Institute, Los Angeles, CA 65–7.

Bound, M., 1998, 'British sea power: ships, armament, strategy and tactics', in Bound, M., ed., *Excavating Ships of War*, Anthony Nelson, Oswestry, 8–30.

Bowens, A., ed., 2009, *Underwater Archaeology: the NAS guide to principles and practice*, Nautical Archaeology Society, Portsmouth, Blackwell Publishing, Oxford.

Bratten, J.R. and Hunter J.W., 1998, 'Conservation', in Smith, R.C., Bratten, J.R., Cozzi, J., and Plaskett, K., *The Emanuel Point Ship: Archaeological Investigations, 1997–1998*, University of West Florida, Pensacola, FL (http://www.flheritage.com/archaeology/projects/shipwrecks/emanuelpoint/epsAI9798.pdf), 158–63.

British Archaeological Association, 1847, 'Proceedings of the Association, December 9th', *Journal of the British Archaeological Association*, **2**, 361–2.

Bowsher, J., 1998, *The Rose Theatre: an archaeological discovery*, Museum of London, London.

Broodbank, J.G., 1921, *History of the Port of London*, Daniel O'Connor, London.

Brooke, C.N.L., and Kier, G., 1975, *London 800–1216: the shaping of a city*, Secker and Warburg, London.

Brooks, G.E., 2003, *Eurafricans in Western Africa: commerce, social status, gender, and religious observance from the 16th to the 18th Century*. Ohio University Press, Athens, OH; James Currey, Oxford.

Brown, R.R., 1997, 'Arms and Armour from Wrecks: an introduction', in Redknap, M., ed., *Artefacts from Wrecks: dated assemblages from the late Middle Ages to the Industrial Revolution*, Oxbow Books, Oxford, 101–10.

Bryson, A., 2008, *The Legal Quays: Sir William Paulet, First Marquis of Winchester*, Lecture for Gresham College, delivered 26 November 2008 (http://www.gresham.ac.uk/lectures-and-events/the-legal-quays-sir-william-paulet-first-marquis-of-winchester).

Buchwald, V. F., 2005, *Iron and steel in ancient times*, det Kongelige Danske Videnskabernes Selskab, Copenhagen.

Buchwald, V.F. and Wivel, H., 1998, 'Slag analysis as a method for the characterization and provenancing of ancient iron objects', *Materials Characterization*, **40(2)**, 73–96.

Butler, R., Green, C. and Payne, N., 2009, 'Cast copper-alloy cooking vessels', *Finds Research Group AD 700–1700 Datasheet*, **41**, 1–9.

Calissendorff. K., 1979, 'Linguistic Evidence for Early Iron Production', in Clarke H, ed. and trans., K. Calissendorff, *et al.*, *Iron and Man in Prehistoric Sweden*, Jernkontoret, Stockholm, 157–75.

Callegari, D., 1994, *Cook's Cannon and Anchors: the recovery and conservation of relics from H.M.B. Endeavour*. Kangaroo Press, Kenthurst, NSW.

Cameron, E., Spriggs, J., and Wills, B., 2006, 'The conservation of archaeological leather', in Kite, M., and Thomson, R., eds, *Conservation of Leather and Related Materials, Butterworth-Heinemann*, Oxford, 244–63.

Caple, C., 2000, *Conservation Skills: Judgement, Method and Decision-Making*, Routledge, London.

Carlin, M, 1996, *Medieval Southwark*, Hambledon Press, London.

Carlin, W. and Keith, D.H., 1996, 'An improved tannin-based corrosion inhibitor-coating system for ferrous artefacts', *International Journal of Nautical Archaeology*, **25(1)**, 38–45.

Carlson, I.M., 2005, *Footwear of the Middle Ages* (http://www.personal.utulsa.edu/~marc-carlson/shoe/SHOEHOM1.HTM)

Carpenter, J., 1990, 'A review of physical methods used to remove concretions from artefacts, mainly iron, including some recent ideas', *Bulletin of the Australian Institute for Maritime Archaeology*, **14(1)**, 25–42.

Carpenter, J. and MacLeod, I.D., 1993, 'Conservation of corroded iron cannon and the influence of degradation on treatment times', in Bridgland, J., ed., *ICOM 10th Triennial Meeting, Washington D.C., USA, 22-27 August 1993, Preprints, vol. 2*, ICOM Committee for Conservation, Paris, 759–66.

Catling, D. and Grayson, J., 1982, *Identification of Vegetable Fibres*, Chapman and Hall, London.

Cederlund, C.O., 1995, 'Marine archaeology in society and science', *International Journal of Nautical Archaeology*, **24(1)**, 9–13.

Charlton, M.F., Crew, P., Rehren, T., and Shennan, S.J., 2010, 'Explaining the evolution of ironmaking recipes – an example from northwest Wales', *Journal of Anthropological Archaeology*, **29(3)**, 352–67.

Charlton, M.F., Blakelock, E., Martinón-Torres, M., and Young, T., 2012, 'Investigating the production provenance of iron artifacts with multivariate methods', *Journal of Archaeological Science*, **39(7)**, 2280–93.

Chirikure, S., Heimann, R. and Killick, D.J., 2010a, 'The technology of tin smelting in the Rooiberg Valley, Limpopo Province, South Africa, *c*. 1650–1850 CE', *Journal of Archaeological Science*, **37(7)**, 1656–69.

Chirikure, S., Sinamai, A., Goagoses, E., Mubusisi, M., and Ndoro, W., 2010b, 'Maritime Archaeology and Trans-Oceanic Trade: a case study of the Oranjemund shipwreck cargo, Namibia', *Journal of Maritime Archaeology*, **5(1)**, 37–55.

Clavir, M., 2009, 'Conservation and Cultural Significance', in Richmond, A., and Bracker, A., eds, *Conservation: principles, dilemmas and uncomfortable truths*, Butterworth-Heinemann, Oxford, 139–49.

Clayton, R.E, Gale, N.H. and Stos-Gale, Z.A., 2002, 'Isotope Fingerprinting of Ancient Tin: preliminary studies', in Jerem, E., and Biro, K.T., eds, *Archaeometry 98: Proceedings of the 31st Symposium, Budapest, April 26 - May 3 1998*, BAR International Series 1043(II), 301–9.

Cleere, H. and Crossley, D., 1995, *The Iron Industry of the Weald (2nd ed.)*, Merton Priory Press, Cardiff.

Colvin, H.M., 1963, *The History of the King's Works: vol. 1*, HMSO, London.

Colvin, H.M. and Foister, S. eds, 1996, *The Panorama of London circa 1544 by Anthonis van den Wyngaerde*, London Topographical Society, London.

Cook, C. and Grattan, D., 1991, 'A method of calculating the concentration of PEG for freeze-drying waterlogged wood', in Hoffman, P., ed., *Proceedings of the 4th ICOM Group on Wet Organic Archaeological Materials Conference, Bremerhaven, 1990*, Deutsches Schiffahrtsmuseum, Bremerhaven, 239–51.

Courtney, T.W., 1974, 'Excavations at the Royal Dockyard, Woolwich, 1972–3, Part', *Post Medieval Archaeology*, **8**, 1–28.

Coustures, M.P., Béziat, D., Tollon, F., Domergue, C., Long, L., and Rebiscoul, A., 2003, 'The use of trace element analysis of entrapped slag inclusions to establish ore-bar links: examples from two Gallo-Roman iron-making sites in France (Les Martys, Montagne Noire, and Les Ferrys, Loiret)', *Archaeometry*, **45(4)**, 599–613.

Cox, N. and Dannehl, K., 2007, *Dictionary of Traded Goods and Commodities, 1550–1820*, University of Wolverhampton (http://www.british-history.ac.uk/report.aspx?compid=58908).

Crew, P., 2000, 'The influence of clay and charcoal ash on bloomery slags', in Tizzoni, C.C., and Tizzoni, M., eds, *Iron in the Alps: Deposits, Mines and Metallurgy from Antiquity to the XVI Century*, Bienno, 38–48.

Croft, P.J., 1973 *The Spanish Company*, London Record Society 7, London.

Cronström, A., 1982, '*Von Eisengruben auf zu finden, des Malms Art und Natur zu kennen und selbiges widerum zu arbeiten*', G. ed., *Iron and Steel on the European Market in the 17th Century: a contemporary Swedish account of production forms and marketing*, The Historical Metallurgy Group of the Swedish Ironmasters' Association, Stockholm, 55–239.

Cronyn, J.M., 1990, *The Elements of Archaeological Conservation*, Routledge, London.

Crossley, D., 1984, 'The survival of early blast-furnaces: a world survey', *Historical Metallurgy*, **18**(2), 112–31.

Cummins, J., 1995, *Francis Drake: the lives of a hero*, Weidenfeld & Nicolson, London.

Degrigny, C. and Spiteri, L., 2004, 'Electrochemical monitoring of marine iron artefacts during their storage/stabilization in alkaline solutions', in Ashton, J., and Hallam, D., eds, *Metal 04: proceedings of the International Conference on Metals Conservation, Canberra, Australia 4-8 October 2004*, National Museum of Australia, Canberra, 315–31.

Degryse, P., Schneider, J.C., Kellens, N., Waelkens, M., and Muchez, P., 2007, 'Tracing the resources of iron working at ancient Sagalassos (south-west Turkey): combined lead and strontium isotope study on iron artifacts and ores', *Archaeometry*, **49(1)**, 75–86.

Degryse, P., Schneider, J.C., and Muchez, P., 2009, 'Combined Pb-Sr isotopic analysis in provenancing late Roman iron raw materials in the territory of Sagalassos (SW Turkey)', *Archaeological and Anthropological Sciences*, **1(3)**, 155–9.

Denslagen, W., 2003, 'The Artificial Life of Heritage', in Theophile, E., and Gutschow, N., eds, *The Sulima Pagoda: East meets West in the restoration of a Nepalese temple*, Weatherhill, MN, 95–102.

Desaulty, A.-M., Dillmann, P., L'Héritier, M., Mariet, C., Gratuze, B., Joron, J. L., and Fluzin, P., 2009, 'Does it come from the Pays de Bray? Examination of an origin hypothesis for the ferrous reinforcements used in French medieval churches using major and trace element analyses', *Journal of Archaeological Science*, **36(10)**, 2445–62.

Devos, W., Senn-Luder, M., Moor, C., and Salter, C., 2000, 'Laser ablation inductively coupled plasma mass spectrometry (LA-ICP-MS) for spatially resolved trace analysis of early-medieval archaeological iron finds', *Fresenius Journal of Analytical Chemistry*, **366(8)**, 873–80.

Dietz, B., 1972, *The Port and Trade of Early Elizabethan London: documents*, London Record Society 8, London.

Dietz, B., 1986, 'Overseas Trade and Metropolitan Growth', in Beier, A.L., and Finlay, R., eds, 1986, *The Making of the Metropolis: London 1500–1700*, Longman, London, 115–40.

Dietz, B., 1991, 'The Royal Bounty and English Shipping in the 16th and 17th centuries', *Mariner's Mirror*, **77(1)**, 5–20.

Dillman, P. and L'Héritier, M., 2007, 'Slag inclusion analyses for studying ferrous alloys employed in French medieval buildings: supply of materials and diffusion of smelting processes', *Journal of Archaeological Science*, **34(11)**, 1810–23.

Domoney, K., 2009, *Gresham Ship, Princes Channel, Thames Estuary: Conservation of an Elizabethan Shipwreck Assemblage*, English Heritage Research Department Report Series 102-2009, English Heritage, Swindon.

Dyson, T., 1989, *Documents & Archaeology: the medieval waterfront*, Museum of London, London.

Eddison, J., 2013, *Medieval Pirates: pirates, raiders and privateers, 1204–1453*, The History Press, Stroud.

Edward de Vere Newsletter, April 1989, February 2001, **2** (http://www.oxford-shakespeare.com/Newsletters/Voyages_of_Explor-02.pdf).

Egan, G., 2005, *Material Culture in London in an age of transition: Tudor and Stuart period finds c. 1450–c.1700 from excavations at riverside sites in Southwark*, Museum of London Archaeology Service monograph 19, London.

Englund, L-E., 2011, 'Sources of Knowledge of Bloomery Iron', in af Geijerstam, J. and Nisser, M., eds, *Swedish Mining and Metalworking – Past and Present,* National Atlas of Sweden, Jernkontoret, Stockholm, 36–7.

Epstein, M., 1908, *The Early History of the Levant Company*, Routledge, London.

Evans, N. and Mould, Q., 2005, 'Footwear', in Gardiner, J., and Allen, M.J., eds, *Before the Mast: Life and Death aboard the Mary Rose*, Mary Rose Trust, Portsmouth, 59–94.

Evans, C. and Rydén, G., 2007, *Baltic Iron in the Atlantic World in the Eighteenth Century*, Brill, Leiden and Boston, MA.

Fenwick, V. and Gale, A., 1998, *Historic Shipwrecks: discovered, protected and investigated*, Tempus, Stroud.

Ffoulkes, C.J., 1969, *The Gun-founders of England (2nd edition)*, Arms and Armour Press, London.

Firth, A., 2006, 'Old Shipwrecks and New Dredging: an Elizabethan ship in the Thames', in Grenier, R., Nutley, D., and Cochran, I., eds, *Underwater Cultural Heritage at Risk*, ICOMOS, Paris, 35–8.

Firth, A., 2011, 'London Gateway: a maritime history', *Talk of the Thames*, Summer 2011, 14–17.

Fisher, F., 1990, *London and the English Economy: 1550–1700*, Hambledon Press, London.

Florian, M.-L.E., 1987, 'The underwater environment', in Pearson, C., ed., *Conservation of Marine Archaeological Objects*, Butterworth, London, 1–20.

Florian, M.-L.E., Kronkright, D.P., Norton, R.E., 1990, *The Conservation of Artifacts made from plant materials*, J. Paul Getty Trust, Malibu, CA.

Forsyth, H., 2003, *The Cheapside Hoard*, Museum of London, London.

Foster, W., 1933 *England's Quest of Eastern Trade*, Black, London.

Friel, I., 1995, *The Good Ship: ships, shipbuilding and technology in England 1200–1520*, British Museum Press, London.

Friel, I., 2003, *Maritime History of Britain and Ireland c. 400–2001*, British Museum Press, London.

Froude, J.A., 1923, *English Seamen in the 16th Century*, Longman, Green and Co., London.

Gadd, D. and Dyson, T., 1981, 'Bridewell Palace Excavations at 9–11 Bridewell Place and 1-3 Tudor Street, City of London, 1978', *Post-Medieval Archaeology*, **15(1)**, 1–79.

Gaimster, D. and Stamper, P., (eds.) 1997, *The age of transition : the archaeology of English culture 1400-1600 : proceedings of a conference hosted by the Society for Medieval Archaeology and the Society for Post-Medieval Archaeology at the British Museum, London, 14–15 November, 1996*, Oxbow monograph series 15, Oxford.

Gale, N.H. and Stos-Gale, Z.A., 2000, 'Lead isotope analyses applied to provenance studies', in Ciliberto, E., and Spoto, G., eds, *Modern Analytical Methods in Art and Archaeology*, Wiley, New York, 503–84.

Ganiaris, H., Keene, S. and Starling, K., 1982, 'A comparison of some treatments for excavated leather', *The Conservator* **6**, 12–23.

Ganiaris, H. and Starling, K., 1996, 'Up to standard: planning the needs of an archaeological archive', in Roy, A., and Smith, P., eds, *Archaeological Conservation and its Consequences: preprints of the contributions to the Copenhagen Congress 26–30 August 1996*, International Institute of Conservation, London, 55–8.

Gardiner, J. and Allen, M. eds, 2005, *Before the Mast: Life and Death aboard the Mary Rose*, Mary Rose Trust, Portsmouth .

Garside, P. and Wyeth, P., 2006, 'Identification of cellulosic fibres by FTIR spectroscopy: differentiation of flax and hemp by polarized ATR FTIR', *Studies in Conservation*, **51(3)**, 205–11.

Gell, A., 1998, *Art and Agency: towards an anthropological theory*, Clarendon Press, Oxford.

Gibbins, D. and Adams, J., 2000, 'Shipwrecks and maritime archaeology', *World Archaeology*, **32(3)**, 279–91.

Gil, P.P., Yusta, I., Orue-Etxebarria, X., Herrero, J.M., Baceta, J.I., and Artaraz, J., Madina, E., and Mintegui, J.M., 2008, 'Caracterización Mineralógica y Química de Escorias de Ferrerías de Monte ("haizeolas") en el Entorno de Galdakao (Vizcaya)', *Revista de la Sociedad Española de Mineralogía*, **9**, 117–18.

Gömöri, J., 1986, 'Spread of blast furnaces in post-medieval Hungary: production, trade and use of iron', in *Medieval Iron in Society II: papers and discussions at the symposium in Norberg, May 6–10, 1985*, Jernkontoret and Riksantikvarieämbetet, Stockholm, 229–39.

Gonzalez, N., Cook, D., de Vivies, P., Drews, M. J. and Mardikian, P., 2007, 'The effects of cathodic polarization, soaking in alkaline solutions and subcritical water on cast iron corrosion products', in Degrigny, C., Van Langh, R., Joosten, I., and Ankersmit, B., eds, *Metal 07: Interim Meeting of the ICOM-CC Metal Working Group Amsterdam, 17–21 September 2007, vol. 3: Use of Electrochemical Techniques in Metal Conservation*, Rijksmuseum, Amsterdam, 32–7.

Goodburn, D., 2002, 'Western Barges', in Ayres, J. and Wroe-Brown, R., *The London Millennium Bridge: excavation of the medieval and later waterfronts at Peter's Hill, City of London, and Bankside, Southwark*, Museum of London Archaeology Service Archaeological Studies series 6, London, 84–9.

Goodman, E. and Barnard, J., 2005, 'Displaying waterlogged archaeological wood at the Museum of London', in Hoffmann, P., ed., *Proceedings of the 9th ICOM Group on Wet Organic Archaeological Materials Conference: Copenhagen 2004*, Hauschildl, Bremerhaven, 109–18.

Gordon, R.B., 1997, 'Process Deduced From Ironmaking Wastes and Artefacts', *Journal of Archaeological Science*, **24(1)**, 9–18.

Goubitz, O., van Driel-Murray, C., and Groenman-van Waateringe, W., 2001, *Stepping Through Time: Archaeological footwear from prehistoric times until 1800*, Stichting Promotie Archeologie, Zwolle, Netherlands.

Gould, R.A.,, 1983, 'Looking below the surface: shipwreck archaeology as anthropology', in Gould, R.A.,, ed., *Shipwreck Anthropology*, University of New Mexico, Albuquerque, NM, 3–22.

Grainger, I. and Hawkins, D., 1988, 'Excavations at the Royal Mint Site: 1986–1988', *London Archaeologist*, **5(16)**, 429–36.

Grainger, I. and Phillpotts, C., 2010, *The Royal Navy victualling yard, East Smithfield, London*, Museum of London Archaeology Service monograph 45, London.

Grant, M.R., 1990, 'Trace elements in southern African artefact tins using NAA', in Pernicka, E., and Wagner, G.A., eds, *Archaeometry '90*, Birkhäuser Verlag, Basel, 165–72.

Grattan, D.W. and Clarke, R.W., 1987, 'Conservation of Waterlogged Wood', in Pearson, C., ed., *Conservation of Marine Archaeological Objects*, Butterworth, London, 164–206.

Greenhill, B., ed., 1995, *The Evolution of the Sailing Ship 1250–1580*, Conway Maritime Press, London.

Grew, F. and de Neergaard, M., 1988, *Medieval Finds from London, 2: Shoes and Pattens*, HMSO, London.

Hakluyt, R. 1599, *The Second Volume of the Principal Navigations, Voyages, Traffiques and Discoveries of the English Nation ....* , George Bishop *et al.*, London.

Hamilton, D. L., 2010, *Methods of conserving archaeological material from underwater sites*. Texas A&M University (nautarch.tamu.edu/CRL/conservationmanual/)

Harris, G.G., 1969, *The Trinity House of Deptford 1514–1660*, Athlone Press, London.

Harris, G.G., ed., 1983, *The Trinity House of Deptford: transactions 1609–35,* London Record Society 19, London.

Hather, J.G., 2000, *The Identification of the Northern European Woods: a guide for archaeologists and conservators*, Archetype Publications, London.

Hawthorne, W., 1999, 'The production of slaves where there was no state: the Guinea-Bissau region 1450–1815', *Slavery and Abolition: Journal of Slave and Post-Slave Studies* **20(2)**, 97–124.

Hayman, R., 2005, *Ironmaking: the history and archaeology of the iron industry*, Tempus, Stroud.

Hedges, R.E.M. and Salter, C.J., 1979, 'Source determination of iron currency bars through analysis of slag inclusions', *Archaeometry*, **21(2)**, 161–75.

Hildebrand, K-G., 1992, *Swedish Iron in the Seventeenth and Eighteenth Centuries: export industry before the industrialization*, Jernkontoret: Stockholm.

Hildred, A., 1997, 'The material culture of the *Mary Rose* (1545) as a fighting vessel: the uses of wood', in Redknap, M., ed., *Artefacts from Wrecks: dated assemblages from the late Middle Ages to the Industrial Revolution*, Oxbow Books, Oxford, 51–72.

Hoover, H.C. and Hoover, H.L., trans., 1950, *Georgius Agricola: De Re Metallica. Translated from the First Latin Edition of 1556*, Dover Publications, New York.

Hovmand, I., 2002, 'Investigations into the mineral content of waterlogged archaeological leather', in Hoffmann, P., Spriggs, J.A., Grant, T., Cook, C., and Recht, A., eds, *Proceedings of the 8th ICOM Group on Wet Organic Archaeological Materials Conference: Stockholm, 11–15 June 2001*, Deutsches Schiffahrtsmuseum, Bremerhaven, 429–38.

Hovmand, I. and Jones, J., 2001, 'Experimental Work on the Mineral Content of Archaeological Leather', in Wills, B., ed., *Leather Wet and Dry: current treatments of waterlogged and desiccated archaeological leather*, Archaeological Leather Group, London, 27–36.

Huda, K., 2002, 'A note on the efficacy of ethylenediaminetetra-acetic acid disodium salt as a stripping agent for corrosion products of copper', *Studies in Conservation*, **47(3)**, 211–16.

Hudgins, C.C., Martinón-Torres, M. and Rehren, T., 2009, 'From the Mines to the Colonies: archaeological evidence for the exchange and metallurgical usage of English copper in early seventeenth-century Ireland and Virginia', in Horning, A., and Brannon, N., (eds.), *Ireland and Britain in the Atlantic World*: 157–80, Wordwell, Dublin.

Huisman, D.J., ed., 2009, *Degradation of Archaeological Remains*, Cultural Heritage Agency/Sdu Uitgevers, Den Haag.

Humpherus, H., 1859, (rep 1999) *History of the Origins and Progress of the Company of Watermen and Lightermen of the River Thames: 1: 1514–1859.* Company of Watermen and Lightermen, London.

Ingason, G., Firth, A., Callan, N., Card, K., and Nichols, K., eds, 2010, *London Gateway: a maritime history*, DP World/Wessex Archaeology, London.

Institute for Archaeologists, 2009, *Standard and Guidance for the creation, compilation, transfer and deposition of archaeological archives*. Reading.

Jackson, G., 2000, 'State concerns for the fisheries 1485–1815', in Starkey, D., Reid, C., and N., Ashcroft, N., eds, *England's Sea Fisheries: the commercial sea fisheries of England and Wales*, Chatham, London, 46–53.

Jakes, K.A., and Mitchell, J.C., 1992, 'The recovery and drying of textiles from a deep ocean historic shipwreck', *The Journal of the American Institute of Conservation*, **31(3)**, 343–53.

Jarvis, R.C., 1976, 'The early customs and custom houses in the Port of London, *Transactions of the London and Middlesex Archaeological Society*, **27**, 271–79.

Jennsen, V. 1987, 'Conservation of wet organic artefacts excluding wood', in Pearson, C., ed., *Conservation of Marine Archaeological Objects*, Butterworth, London, 122–63.

Jobling, J., 1990, 'The cast iron cannon from Chandeleur Islands', in Carrell, T.L., (ed), *Underwater Archaeology: proceedings from the Society for Historical Archaeology conference: Tuscon, Arizona 1990,* Society for Historical Archaeology, Pleasant Hill, CA, 97–100.

Jockenhövel, A. and Willms, C., 1997, 'Archaeological investigations on the beginning of blast-furnace technology in central Europe', in Crew, P., and Crew, S., (eds.), *Early Ironworking in Europe: archaeology and experiment: abstracts of the international conference at Plas Tan y Bwlch 19–25 Sept. 1997*, Snowdonia National Park Study Centre, Plas Tan y Bwlch, 56–8.

Joint Iron Council, 1965, *A Guide to the Engineering Properties of Iron Castings (2nd Edition)*, London.

Jones, E., 2009, *'Particular' Accounts and Port Books of the Sixteenth Century, 1503–1601 (Computer File). Colchester, Essex: UK Data Archive* (http://dx.doi.org/10.5255/UKDA-SN-6275-1).

Jones, M., ed., 2003, *For Future Generations: conservation of a Tudor maritime collection*, Mary Rose Trust, Portsmouth .

Joosten, I., 2004, *Technology of Early Historical Iron Production in the Netherlands*, Institute for Geo- and Bioarchaeolgy, Vrije Universiteit, Amsterdam.

Karsten, A. and Graham, K., 2011, *Leather Drying Trial: a comparative study to evaluate different treatment and drying techniques for wet, archaeological leather*, English Heritage Research Report Series, no 70-2011, Swindon.

Karsten, A. Graham, K., Jones, J., Mould, Q. and Walton Rogers, P, 2012, *Waterlogged Organic Artefacts: guidelines on their recovery, analysis and conservation*, English Heritage, Swindon.

Kelsey, H., 2003, *Sir John Hawkins: Queen Elizabeth's slave trader*, Yale University Press, New Haven, CT.

Kelso, W.M., Lucckellt, N.M., and Straube, B.A., 1997, *Jamestown Rediscovery III*, Association for the Preservation of Virginia Antiquities, Richmond VA.

Kennard, A.N., 1986, *Gunfounding and Gun Founders: a directory of cannon founders from earliest times to 1850*, Arms and Armour Press, London

Kiernan, D., 1989, *The Derbyshire lead industry in the sixteenth century*, Derbyshire Record Society in association with the Peak District Mines Historical Society and the Derbyshire Archaeological Society, Chesterfield.

Killock, D., Meddens, F., Armitage, P., Egan, G., Gaimster, D., Jarrett, C., Keys, L., Phillpotts, C., Sabel, K., Tyson, R., and Willmott, H., 2005, 'Pottery as plunder: a 17th-century maritime site in Limehouse, London', *Post-Medieval Archaeology*, **39(1)**, 1–91.

King, P.W., 2004, *The iron trade in England and Wales, 1500-1815: the charcoal iron industry and its transition to coke*, unpublished doctoral thesis, University of Wolverhampton.

King. P.W, 2005, 'The production and consumption of bar iron in early modern England and Wales', *Economic History Review* **58(1)**, 1–33.

Kite, M. and Thomson, R., 2006, *Conservation of Leather and Related Materials*, Butterworth-Heinemann, Oxford.

Kopytoff, I., 1986, 'The cultural biography of things: commoditization as process', in Appadurai, A., ed., *The Social Life of Things: commodities in cultural perspective*, Cambridge University Press, Cambridge, 64-91.

Kowaleski, M, 2000 'The internal fish trade', in Starkey, D., Reid, C., and N., Ashcroft, N., eds, *England's Sea Fisheries: the commercial sea fisheries of England and Wales*, Chatham, London, 29–32.

Krop, D.S. and Nordgren, E., 2011, 'Disassembly of USS *Monitor*'s complex mechanical components', in Mardikian, P., Chemello, C., Watters, C., and Hull, P., eds, *Metal 2010, proceedings of the interim meeting of the ICOM-CC Metal Working Group, 11–15 October 2010, Charleston, South Carolina, USA*, Clemson University, Charleston, SC, 70–8.

Kuleff, I. and Pernicka, E., 1995, 'Instrumental neutron activation analysis of native copper: some methodological considerations', *Journal of Radioanalytical and Nuclear Chemistry*, **191(1)**, 145–61.

Kuznetsov, Y.I., 1996, *Organic Inhibitors of Corrosion of Metals*, Plenum Press, London and New York.

Ladle, L., 1993, *The Studland Bay Wreck*, Poole Museum Service, Poole.

Landi, S., 1998, *The Textile Conservators Manual: Revised Second Edition*, Butterworth-Heinemann, Oxford.

Lang, J., 2005, 'Metals', in Lang, J., and Middleton, A., eds, *Radiography of Cultural Material (2nd edition)*, Butterworth-Heinemann, Oxford, 49–75.

Lang, R.G., 1993, *Two Tudor Subsidy Assessment Rolls for the City of London*, London Record Society 29, London.

Larsen, E.B., 1981, *Moulding and Casting of Museum Objects*, Det Kongelige Danske Kunstakademi, Copenhagen.

Laughton, J.K., ed., 1981, *State Papers relating to the defeat of the Spanish Armada, vol. 2 (2nd edition)*, Navy Records Society, London.

Lenihan, D.J., 1983, 'Rethinking shipwreck archaeology: a history of ideas and considerations for new directions', in Gould, R.A.,, ed., *Shipwreck Anthropology*, University of New Mexico, Albuquerque, NM, 37–64.

Leroy, S., Cohen, S.X., Verna, C., Gratuze, B., Téreygeol, F., Fluzin, P., and Dillmann, P., 2012, 'The medieval iron market in Ariège (France). Multidisciplinary analytical approach and multivariate analyses', *Journal of Archaeological Science*, **39(4)**, 1080–93.

L'Héritier, M., Dillman, P. and Benoit, P., 2010, 'Iron in the building of gothic churches: its role, origins and production using evidence from Rouen and Troyes', *Historical Metallurgy*, **44(1)**, 21–35.

Loades, D., 2009, *The Making of the Elizabethan Navy 1540–1590*, Boydell, Woodbridge.

Loengard, J., 1989, *London Viewers and their Certificates 1508–1558*, London Record Society 26, London.

Maarleveld, T., and Overmeer, A., 2013, Aanloop Molengat – maritime archaeology and intermediate trade during the Thirty Years' War, *Journal of Archaeology in the Low Countries*, **4(1)**, 95–149.

McCarthy, M., 2005, *Ships' Fastenings: from sewn boat to steamship*, Texas A&M University Press, College Station, TX.

McDermott, J., 2001, *Martin Frobisher: Elizabethan Privateer*, Yale University Press, New Haven, CT.

McDermott, J., 2005, *England and the Spanish Armada: the Necessary Quarrel*, Yale University Press, New Haven, CT .

McDonnell, K.G.T., 1978, *Medieval London Suburbs*, Phillimore, London.

Mackenzie, R.J., and Whiteman, J.A., 2006, 'Why pay more? An archaeometallurgical investigation of 19th-century Swedish wrought iron and Sheffield blister steel', *Historical Metallurgy*, **40(2)**, 138–49.

Mackinder, A., and Blatherwick, S., 2000 *Bankside: excavations at Benbow House Southwark, London SE1*, Museum of London Archaeology Service, Archaeology Studies 3, London.

MacLeod, I.D., 1982, 'Formation of marine concretions on copper and its alloys', *International Journal of Nautical Archaeology*, **11(4)**, 267–75.

MacLeod, I.D., 1987, 'Conservation of corroded iron artefacts – new methods for on-site preservation and cryogenic deconcreting', *International Journal of Nautical Archaeology* **16(1)**, 49–56.

MacLeod, I.D., 1991, 'Identification of corrosion products on non-ferrous metal artefacts recovered from shipwrecks', *Studies in Conservation*, **36(4)**, 222–34.

MacLeod, I.D., 1994, 'Conservation of corroded metals: a study of ships' fastenings from the wreck of HMS *Sirius* (1790)', in Scott, D.A., Podany, J., and Considine, B.B., eds, *Ancient & Historic Metals: conservation and scientific research: proceedings of a symposium organized by the J. Paul Getty Museum and the Getty Conservation Institute, November 1991*, Getty Conservation Institute, Los Angeles, CA, 265–78.

MacLeod, I.D, and Flecker, M., 2001, 'Corrosion of tin and its alloys recovered from a 10th century wreck in the Java Sea', in MacLeod, I.D., Theile, J.M., and Degrigny, C., eds, *METAL 2001: proceedings of the international conference of metals conservation: Santiago, Chile, 2–6 April 2001*, Western Australian Museum, Perth, 191–201.

MacLeod, I.D., Mardikian, P., and Richards, V., 1994, 'Observations on the extraction of iron and chloride from composite materials', in Hoffman, P., Daley, T., and Grant, T., eds, *Proceedings of the 5th ICOM Group on Wet Organic Archaeological Materials conference, Portland/Maine, 1993*, Deutsches Schiffahrtsmuseum, Bremerhaven, 199–211.

MacLeod, I.D., and Richards, V.L, 1997, 'The impact of metal corrosion products on the degradation of waterlogged wood recovered from historic shipwreck sites', in Hoffman, P., Grant, T., and Spriggs, J.A., *Proceedings of the 6th ICOM Group on Wet Organic Archaeological Materials Conference, York, 9–13 September 1996*, Deutsches Schiffahrtsmuseum, Bremerhaven, 331–49.

MacLeod, I.D., and Wozniak, R., 1997, 'Corrosion and Conservation of Tin and Pewter from Seawater', in MacLeod, I.D., Pennec, S.L., and Robbiola, L., eds, *Metals 95: proceedings of the International Conference on Metals Conservation, Semur-en-Auxois, 25–28 September 1995*, James & James, London, 118–23.

Magnus, O., 1555, *Historia de Gentibus Septentrionalibus*, Rome.

Magnusson, G., 1995, 'Iron production, Smithing and Iron Trade in the Baltic during the late Iron Age and Early Middle Ages (*c.* 5th–13th Centuries)', in Jansson, I., ed., *Archaeology East and West of the Baltic: papers from the Second Estonian-Swedish Archaeological Symposium, Sigtuna, May 1991*, University of Stockholm, Stockholm, 61–70.

Magnusson, G., 2011a, 'Iron Production in Europe', in af Geijerstam, J., and Nisser, M., eds, *Swedish Mining and Metalworking: past and present*, National Atlas of Sweden, Stockholm, 38–9

Magnusson, G., 2011b, 'The Inception of the Blast Furnace', in af Geijerstam, J., and Nisser, M., eds, *Swedish Mining and Metalworking: past and present*, National Atlas of Sweden, Stockholm, 42–3.

Malcolm, C., 2003, *Trade Goods on the* Henrietta Marie *and the Price of Men in 1699-1700*, Historical Archaeology Conference on Underwater Archaeology 1998 (http://www.melfisher.org/pdf/The%20Price%20of%20Men%20in%201699-1700_Updated-Version.pdf).

Mardikian, P. and David, R., 1996, 'Conservation of a French pistol from the wreck of *Le Cygne* (1808)', *Studies in Conservation*, **41(3)**, 161–9.

Mardikian, P., Chemello, C., Watters, C., and Hull, P., eds, 2010, *Metal 2010, proceedings of the interim meeting of the ICOM-CC Metal Working Group, 11–15 October 2010, Charleston, South Carolina, USA*, Clemson University, Charleston, SC, 70–8.

Marken, M.W., 1994, *Pottery from Spanish Shipwrecks, 1500–1800*, University Press of Florida, Gainesville, FL.

Marsden, P., 1996, *Ships of the Port of London: twelfth to seventeenth centuries AD,* English Heritage Archaeological Report 5, London.

Marsden, P., ed., 2003, *Sealed by Time: the loss and recovery of the Mary Rose*, Mary Rose Trust, Portsmouth.

Martin, C., 1972, '*El Gran Grifon*: An Armada wreck on Fair Isle', *International Journal of Nautical Archaeology* **1(1)**, 59–71.

Martin, C., 1997, 'Ships as integrated artefacts: the archaeological potential', in Redknap, M., ed., *Artefacts from Wrecks: dated assemblages from the late Middle Ages to the Industrial Revolution*, Oxbow Books, Oxford, 1–13.

Martin, C., 2000, 'De-particularizing the particular: approaches to the investigation of well-documented post-medieval shipwrecks', *World Archaeology*, **32(3)**, 383–99.

Martin, C., 2005, 'The *Adelaar*: a Dutch East-Indiaman Wrecked in 1728 off Barra, Outer Hebrides, Scotland', *International Journal of Nautical Archaeology*, **34(2)**, 179–210.

Martin, C., and Parker, G., 1988, *The Spanish Armada*, Hamish Hamilton, London.

Mattingly, G., 1959, *The Defeat of the Spanish Armada*, Jonathan Cape, London.

Mills, J.S. and White, R., 1989, The identity of the resins from the late bronze age shipwreck at Ulu Burun (Kas), *Archaeometry*, **31(1)**, 37–44.

Milne, G., 2003, *The Port of Medieval London*, Tempus, Stroud.

Moorhouse, G., 2005, *Great Harry's Navy: how Henry VIII gave England Sea Power*, Weidenfeld & Nicolson, London.

Morelli, R., 1982, 'Swedenborg and an iron foundry in Italy', *Historical Metallurgy*, **16(1)**, 24–8.

Morris, D. and Cozens, K., 2009, *Wapping 1600–1800: a social history of an early modern London maritime suburb*, East London History Society, London.

Muckelroy, K., 1998, 'Introducing maritime archaeology', in Babits, L.E., and Van Tilburg. H., eds, *Maritime Archaeology: a reader of substantive and theoretical contributions*, Plenum Press, London, 23–37.

Muncher, D.A., 1988, 'Composite casting of partially degraded iron artifacts', *Studies in Conservation*, **33(2)**, 94–6

Muñoz Viñas, S., 2005, *Contemporary Theory of Conservation*, Elsevier Butterworth-Heinemann, Oxford and Burlington, MA

Muñoz Viñas, S., 2009, 'Minimal intervention Revisited', in Richmond, A., and Bracker, A., eds, *Conservation: principles, dilemmas and uncomfortable truths*, Butterworth-Heinemann, Oxford, 47–59.

Murphy, L., 1983, 'Shipwrecks as data base for human behavioural studies', in Gould, R.A.,, ed., *Shipwreck Anthropology*, University of New Mexico, Albuquerque, NM, 65-90.

Museum of London Archaeological Archive, 2013, *Standards 2, Archive Components: Standards and Specifications 2.4 Finds Conservation* (http://www.museumoflondon.org.uk/files/6613/7234/1666/Standards_24_Finds_conservation.pdf)

Needham, J., 1986, *Science and Civilization in China: Volume 4 (Physics and Physical Technology) Part 2 (Mechanical Engineering)*, Caves Books, Taipei, Taiwan.

National Maritime Museum, 1998, *Armada 1588–1988*, Penguin Books, London

Noël Hume, I., 1994, *The Virginia Adventure Roanoke to James Towne: an archaeological and historical odyssey*, Knopf, New York.

North, N.A., 1987, 'Conservation of metals', in Pearson, C., ed., *Conservation of Marine Archaeological Objects*, Butterworth, London, 207–52.

North, N.A., and Owens, M., 1981, 'Design and operation of a furnace for H_2 reduction of marine iron', *International Journal of Nautical Archaeology*, **10(2)**, 95–100.

North, N.A., and MacLeod, I.D., 1987, 'Corrosion of metals', in Pearson, C., ed., *Conservation of Marine Archaeological Objects*, Butterworth, London, 68–104.

Nyquist, O., 2011, 'The Geography of Mining and Metalworking', in af Geijerstam, J. and Nisser, M., eds, *Swedish Mining and Metalworking – Past and Present,* National Atlas of Sweden, Jernkontoret, Stockholm 16–17.

Odegaard, N., Carroll, S., and Zimmt, W.S., 2000, *Material characterization tests for objects of art and archaeology*, Archetype Publications, London.

Oppenheim, M., 1896, *A History of the Administration of the Royal Navy, vol. 1, 1509-1660*, John Lane, London.

Oxley, I., 1998, 'The investigation of the factors that affect the preservation of underwater archaeological sites', in Babits, L.E., and Van Tilburg. H., eds, *Maritime Archaeology: a reader of substantive and theoretical contributions*, Plenum Press, London, 523–29.

Pagès, G., Dillman, P., Fluzin, P. and Long, L., 2011, 'A study of the Roman iron bars of Saintes-Maries-de-la-Mer (Bouches-du-Rhône, France). A proposal for a comprehensive metallographic approach', *Journal of Archaeological Science*, **38(6)**, 1234–52.

Panter, I. and Spriggs, J., 1997, 'Condition Assessments and Conservation Strategies for Waterlogged Wood Assemblages', in Hoffman, P., Grant, T., and Spriggs, J.A., *Proceedings of the 6th ICOM Group on Wet Organic Archaeological Materials Conference, York, 9–13 September 1996*, Deutsches Schiffahrtsmuseum, Bremerhaven, 180–200.

Parker, G., 1996, 'The Dreadnought Revolution of Tudor England', *Mariner's Mirror* **82(3)**, 269–300.

Payne, A., 2008, *Richard Hakluyt: a guide to his books and to those associated with him, 1580–1625*, Bernard Quaritch, London.

Paynter, S., 2006, 'Regional variations in bloomery smelting slag of the Iron Age and Romano-British periods', *Archaeometry,* **48(2)**, 271–92.

Pavitt, G., 2009, *Sir Thomas Gresham and the Royal Exchange,* lecture presented at Gresham College London on 16th May 2009.

Peacock, E.E., 1990, Freeze-drying archaeological textiles: the need for basic research, in O'Connor, S.A., and Brooks, M.M., eds, *Archaeological Textiles: the proceedings of the conference, Textiles for the Archaeological Conservator held by the UKIC Archaeology Section, York, April 1988*, United Kingdom Institute for Conservation Occasional Papers 10, London, 22–30.

Peacock, E.E, 2001, 'Water-degraded Archaeological Leather: an overview of treatments used at Vitenskapsmuseum (Trondheim)', in Wills, B., ed., *Leather Wet and Dry: current treatments of waterlogged and desiccated archaeological leather*, Archetype Publications for the Archaeological Leather Group, London, 11–26.

Pearson, C., 1987, 'On-site storage and conservation', in Pearson, C., ed., *Conservation of Marine Archaeological Objects*, Butterworth, London, 105–16.

Pipping, G., 1982, 'Glossary', in Pipping, G. ed., *Iron and Steel on the European Market in the 17th Century: a contemporary Swedish account of production forms and marketing*, The Historical Metallurgy Group of the Swedish Ironmasters' Association, Stockholm, 240–54.

Price, R., Muckleroy, K., and Willies, L., 1980, 'The *Kennemerland* Site: a report on the lead ingots', *International Journal of Nautical Archaeology*, **9(1)**, 7–25.

Pye, E., 2001, *Caring for the past: issues in conservation for archaeology and museums*, James & James, London.

Quennell, P ed., 1984, *Mayhew's London: being selections from "London labour and the London poor"*, Bracken Books, London.

R Core Team. 2012, *R: A Language and Environment for Statistical Computing*, R Foundation for Statistical Computing, Vienna.

Ramsay, G.D., 1986, *The Queen's Merchants and the Revolt of the Netherlands*, Manchester University Press, Manchester.

Reddaway, T.F, 1940, *The Rebuilding of London after the Great Fire*, Jonathan Cape, London.

Redknap, M., 1984, *The Cattewater Wreck: the investigation of an armed vessel of the early sixteenth century*, BAR British Series 131, Oxford.

Redknap, M., 1985, 'The Cattewater Wreck: a contribution to 16th-century maritime archaeology', in Cederlund, C.O., ed., *Postmedieval Boat and Ship Archaeology*, BAR International Series 256, Oxford, 39–47.

Redknap, M., 1997, 'Reconstructing 16th-century ship culture: the Cattewater wreck', in Redknap, M., ed., *Artefacts from Wrecks: dated assemblages from the late Middle Ages to the Industrial Revolution*, Oxbow Books, Oxford, 73–85.

Rees-Jones, S. G., 1972, 'Some aspects of conservation of iron objects from the sea', *Studies in Conservation*, **17(1)**, 39–43.

Reguer, S., Neff, D., Remazeilles, C., Guilminot, E., Nicot, F., Pelé, C., Meguelati, M., Mirambet, F., Dillmann, P., Refait, P., Huet, N., Mielcarek, F., Rebière, J. and Bertrand, L., 2007, 'Desalinisation of iron archaeological artefacts: understanding of chlorine removal mechanisms of the corrosion layers supported by characterisation techniques', in Degrigny, C., Van Langh, R., Joosten, I., and Ankersmit, B., eds, *Metal 07: Interim Meeting of the ICOM-CC Metal Working Group Amsterdam, 17–21 September 2007, vol. 2. Innovative investigation of metal artefacts*, Rijksmuseum, Amsterdam, 60–8.

Reynolds, T.S., 1986, 'Iron and water: technological context and the origins of the water-powered iron mill', in *Medieval Iron in Society II: papers and discussions at the symposium in Norberg, May 6–10, 1985*, Jernkontoret and Riksantikvarieämbetet, Stockholm, 61–80.

Rinman, S, 1788–9, *Bergverks Lexicon*, Johan A. Carlbohm, Stockholm.

Robinson, W., 1998, *First Aid for Underwater Finds*, Archetype Publications, London.

Rohl, B.M., 1996, 'Lead isotope data from the Isotrace Laboratory, Oxford: archaeometry data base 2, galena from Britain and Ireland', *Archaeometry*, **38(1)**, 165–80.

Rosenqvist, A.M., 1983, 'Report on chemical and mineralogical analyses of Norwegian ores, slags and iron', *Offa*, **40**, 129–37.

Rostoker, W. and Dvorak, J., 1990, 'Wrought irons: distinguishing between processes', *Archeomaterials*, **4(2)**, 153–66.

Rule, M., 1982, *The Mary Rose: the excavation and raising of Henry VII's flagship*, Conway Maritime Press, London.

San Claudio Santa Cruz, M., 2008, Prospección arqueolóxica subacuática no pecio de punta Restelos, Cée, Ría de Corcubión (A Coruña), in *Actuacións arqueolóxicas, ano 2006*, Consellería de Cultura, Xunta de Galicia, La Coruña.

Salisbury, W., and Anderson, R.C., eds, 1958, *A Treatise on Shipbuilding and a Treatise on Rigging written about 1620–1625*, Society for Nautical Research, London

Salzman, L.F., 1952, *Building in England down to 1540: a documentary history*, Clarendon Press, Oxford.

Satchell, J., and van Rensburg, J.J., 2006, *Quantifying the Hazardous Threat: an assessment of site monitoring data and environmental data sets. Project Report,* Hampshire & Wight Trust for Maritime Archaeology and English Heritage (https://www.academia.edu/5813046/Quantifying_the_Hazardous_Threat_an_assessment_of_site_monitoring_data_and_environmental_datasets).

Saunders, A.L., ed., 1997, *The Royal Exchange*, London Topographical Society no 152, London.

Scammell, G., 1981, *The World Encompassed: the first European maritime empires* c. *800–1650*, Methuen, London.

Schofield, J., 1994, *Medieval London Houses*, Yale University Press, New Haven, CT.

Schofield, J., 2013, 'The Medieval Port of London: publication and research access', *London Archaeologist*, **13(7)**, 181–6

Scott, D.A., 2002, *Copper and bronze in art: corrosion, colorants, conservation*, Getty Conservation Institute, Los Angeles, CA.

Scott, D.A., and Eggert, G., 2009, *Iron and steel in art: corrosion, colorants, conservation*, Archetype, London

Sharpe, R., 1876, *A short account of the worshipful company of shipwrights.* London.

Smith, C.W., 2003, *Archaeological conservation using polymers. Practical applications for organic arefact stabilization*, Texas A&M University Press, College Station, TX.

Smith, H.A., 1991, Conserving iron objects from shipwrecks: a new approach, in Vandiver, P.A., Druzik, J.R., and Wheeler, G.S., eds, *Materials issues in art and archaeology II: symposium held April 17-21, 1990, San Francisco, California, U.S.A.*, Materials Research Society, Pittsburgh, PA, 761–4.

Smith, R.C., Bratten, J.R., Cozzi, J., and Plaskett, K., *The Emanuel Point Ship: Archaeological Investigations, 1997–1998*, University of West Florida, Pensacola, FL (http://www.flheritage.com/archaeology/projects/shipwrecks/emanuelpoint/epsAI9798.pdf).

Smith, T., 1997, 'The Jernkontoret 250th anniversary conference', *Historical Metallurgy*, **31(1)**, 36–41.

Smith, V.T.C., 1985, *Defending London's River: the story of the Thames forts 1540-1945*, North Kent Books, Rochester.

Spazier, I., 2003, 'The Germanic iron-smelting complex at Wolkenberg in Lower Lausacia, Southern Brandenburg', in Nørbach, L.C., ed., *Prehistoric and Medieval Direct Iron Smelting in Scandinavia and Europe: Aspects of Technology and Society*, Aarhus University Press, Aarhus, 37–41.

Spence, R.T., 1995, *The Privateering Earl: George Clifford, 3rd Earl of Cumberland, 1558–1605*, Allan Sutton, Stroud.

Starley, D., 1999, 'Determining the Technological Origins of Iron and Steel', *Journal of Archaeological Science*, **26(8)**, 1127–33.

Steedman, K., Dyson, T. and Schofield, J., 1992, *Aspects of Saxo-Norman London 3: Bridgehead and Billinmgsgateto 1200*, London and Middlesex Archaeological Society Special Paper no 14, London.

Stos-Gale, Z.A. and Gale, N.H., 2009, 'Metal provenancing using isotopes and the Oxford archaeological lead isotope database (OXALID)', *Archaeological and Anthropological Sciences* **1(3)**, 195–213.

Stos-Gale, Z.A., Gale, N.H., Bass, G., Pulak, C., Galili, E. and Sharvit, J., 1998, 'The copper and tin ingots of the Late Bronze Age Mediterranean: new scientific evidence', in Gakkai, N.K., ed., *Proceedings: The Fourth International Conference on the Beginning of the Use on Metals and Alloys (BUMA-IV): May 25-27, 1998, Kunibiki Messe, Japan*, The Japan Institute of Metals, Aoba, Japan, 115–26..

Sturdy, D., 1977, 'Boat Archaeology of the Thames', in McGrail, S., ed., *Sources and Techniques in Boat Archaeology: papers based on those presented to a symposium at Greenwich in September*

1976, together with edited discussion, BAR Supplementary Series 29, Oxford, 289–97.

Suenson-Taylor, K. and Sully, D., 1997, 'The use of condition score to determine glycerol concentration in the treatment of waterlogged archaeological leather: an experimental solution', in Hoffman, P., Grant, T., and Spriggs, J.A., *Proceedings of the 6th ICOM Group on Wet Organic Archaeological Materials Conference, York, 9–13 September 1996*, Deutsches Schiffahrtsmuseum, Bremerhaven, 157–72.

Sully, D., 2014, 'Conservation Theory and Practice: materials, values, and people in heritage conservation', in McCarthy, C., ed. *Museum Practice: Critical Debates in the Museum Sector*, International Handbooks of Museum Studies Volume 4, Wiley Blackwell, Oxford, 1–23.

Sully, D., Domoney, K. and Saunders, J., forthcoming, 'Monitoring the condition of the Gresham Ship Hull', *Studies in Conservation.*

Symons, T.H.B., ed., 1999, *Meta Incognita: a discourse of discovery: Martin Frobisher's Arctic voyages, 1576–1578*, Canadian Museum of Civilization, Hull, Quebec.

Tatton-Brown, T., 1974, 'Excavations at the Custom House Site, City of London', 1973', *Transactions of the London and Middlesex Archaeological Society*, **25**, 117–219.

Tatton-Brown, T., 1975, 'Excavations at the Custom House Site, City of London', 1973 – Part 2', *Transactions of the London and Middlesex Archaeological Society*, **26**, 103-70.

Tatton-Brown, T., 2000, *Lambeth Palace: a history of the Archbishops of Canterbury and their Houses*, Society for Promoting Christian Knowledge, London.

Teasdale, E., 1991, *Gunfounding in the Weald in the sixteenth century*, Royal Armouries monograph 2, Trustees of the Royal Armouries, London.

Thibodeau, A.M., Killick, D.J., Ruiz, J., Chesley, J.T., Deagan, K., Cruxent, J.M., and Lyman, W., 2007, 'The strange case of the earliest silver extraction by European colonists in the New World', *Proceedings of the National Academy of Sciences of the United States*, **104(9)**, 3663–6.

Tholander, E., 1989, 'Microstructure examination of slags as an instrument for identification of ancient iron-making processes', in Pleiner, R., ed., *Archaeometallurgy of Iron: International symposium of the Comité pour la sidérurgie ancienne de l'UISPP: Liblice, 5–9 October 1987*, Archaeological Institute ČSAV, Prague, 35–41.

Tholander, E. and Blomgren, S., 1986, 'Some aspects of the origin of the blast furnace', *Historical Metallurgy*, **20(2)**, 79–85.

Thurley, S., 1999, *Whitehall Palace: an architectural history of Royal apartments, 1240–1690*, Yale University Press, New Haven, CT.

Timberlake, S. and Prag, A.J.N.W., 2005, *The Archaeology of Alderley Edge: survey, excavation and experiment in an ancient mining landscape*, BAR British Series 396, Oxford.

Turgoose, S., 1989, 'Corrosion and structure: modelling the preservation mechanisms', in Janaway, R., and Scott, B., eds, *Evidence Preserved in Corrosion Products: new fields in artifact studies: proceedings of a joint conference between UKIC Archaeology Section and the Council for British Archaeology Science Committee, Leeds 1983*, United Kingdom Institute for Conservation Occasional Papers 8, London, 30–2.

Turner, G.L'E., 2000, *Elizabethan Instrument Makers: the origins of the London trade in precision instrument making*, Oxford, University Press, Oxford.

Tylecote, R.F., 1976, *A History of Metallurgy*, The Metals Society, London.

Tylecote, R.F., 1986, 'The early history of the iron blast furnace in Europe; a case of east-west contact?' in *Medieval Iron in Society II: papers and discussions at the symposium in Norberg, May 6–10, 1985*, Jernkontoret and Riksantikvarieämbetet, Stockholm, 158–80.

Tylecote, R.F., 1987, *The early history of metallurgy in Europe*, Longman, London and New York.

Tylecote, R.F., 1991, 'Iron in the Industrial Revolution', in Tylecote, R.F., and Day, J., eds, *The Industrial Revolution in Metals*, Institute of Metals, London, 200–60.

Unger, R., 1980, *The Ship in the Medieval Economy 600–1600*, Croom Helm, London.

Unwin, G., 1966, *The Gilds and Companies of London (4th edition)*, Methuen, London.

Van der Wee, H., 1963, *The Growth of the Antwerp Market and the European Economy: fourteenth–sixteenth centuries*, Martinus Nijhoff, The Hague.

Van Driel-Murray, C., 1987, 'Roman footwear: a mirror of fashion and society', in Friendship-Taylor, D.E., Swann, J.M., and Thomas, S., eds, *Recent Research in Archaeological Footwear*, Association of Archaeological Illustrators and Surveyors in association with Archaeological Leather Group, London, 32–42.

Van Laun, J., 1979, '17th-century ironmaking in South West Herefordshire', *Historical Metallurgy*, **13(2)**, 55–65.

Venables, W.N., and Ripley, B.D., 2002, *Modern Applied Statistics with S (4th edition)*, Springer, New York.

von Brandenberg, M., 2009, *The making of medieval shoes according to methods from the turn of the 14th century (with modifications for shoes from the 10th to the 15th century)* (http://www.mvbmvb.com/html/manufacture/leather/shoes/makingshoes.doc).

Wang, Q., Dove, S., Shearman, F., and Smirnou, M., 2008, 'Evaluation of methods of chloride ion concentration determination and effectiveness of desalination treatments using sodium hydroxide and alkaline sulphite solutions', *The Conservator*, **31**, 67–74.

Waters, D.W., 1958, *The Art of Navigation in England in Elizabethan and Stuart times*, Hollis & Carter, London.

Waters, D.W., 1967, *The Rutters of the Sea: the sailing directions of Pierre Garcie: a study of the first English and French printed sailing directions*, Yale University Press, New Haven, CT.

Watkinson, D., and Neal, V., 1998, *First Aid for Finds (3rd edition)*, Rescue (Trust), UK Institute for Conservation Archaeology Section) and Museum of London, London.

Watkinson, D., and Al-Zahrani, A., 2008, 'Towards quantified assessment of aqueous chloride extraction methods for archaeological iron: de-oxygenated treatment environments', *The Conservator*, **31**, 75–86.

Watson, B., Brigham T., and Dyson, T., 2001, *London Bridge: 2000 years of a river crossing*, Museum of London Archaeology Service monograph 8, London.

Watson, J., 1985, 'Research into aspects of Freeze-Drying Hardwoods between 1982 and 1984', in *Waterlogged Wood: study and conservation. proceedings of the 2nd ICOM Waterlogged Wood Working Group Conference, Grenoble 28–31 August 1984*, Centre d'Étude et de Traitement des Bois Gorgés d'Eau, Grenoble, 213–18.

Watson, P.J., 1983, 'Method and theory in shipwreck archaeology', in Gould, R.A., ed., *Shipwreck Anthropology*, University of New Mexico, Albuquerque, NM, 23–36.

Weinstein, R., 1994, *Tudor London*, HMSO, London.

Wendrich, W., 1999, *Who is afraid of Basketry: a guide to recording basketry and cordage for Archaeologists and ethnographers*, Centre for Non-Western Studies, Leiden, Netherlands.

Wevers, A.J.M., 1991, 'Treatment of waterlogged rope', in in Hoffman, P., ed., *Proceedings of the 4th ICOM Group on Wet Organic Archaeological Materials Conference, Bremerhaven, 1990*, Deutsches Schiffahrtsmuseum, Bremerhaven, 159–72.

Wheatley, H. ed., 1956, *Stow's Survey of London*, J.M. Dent & Sons, London and New York.

White, R. and Page, H., eds, 1992, *Organic residues in archaeology: their identification and analysis*, UK Institute for Conservation Archaeology Section, London.

Whiting, D, and Willies, L., 1985, 'Lead Ingots', in Larn, R., ed., 'The Wreck of the Dutch East Indiaman *Campen* on the Needles rocks, Isle of Wight, 1627 – part 2', *International Journal of Nautical Archaeology*, **14(2)**, 97–118.

Willan, T.S., 1955, 'Some aspects of English trade with the Levant in the sixteenth century', *English Historical Review*, **70(3)**, 399–410.

Willan, T.S., 1956, *The Early History of the Russia Company: 1553-1603*, Manchester University Press, Manchester.

Williams, A.R., 1990, 'Slag inclusions in armour', *Historical Metallurgy*, **24(2)**, 69–80.

Williams, N., 1955, 'The London Port Books', *Transactions of the London and Middlesex Archaeological Society*, **18**, 13–26.

Wilson, D., 1987, *The Thames: record of a working waterway*, Batsford, London

Weizhen, O. and Chunchun, X., 2005, 'Studies on localized corrosion and desalination treatment of simulated cast iron artefacts', *Studies in Conservation*, **50**, 101–8.

Wood, A., 1999, *The Politics of Social Conflict: the Peak Country, 1520-1770*, Cambridge University Press, Cambridge.

Wouters, J., 1986, 'A comparative investigation of methods for the consolidation of wet archaeological leather: application of freeze-drying to polyethylene glycol impregnated leather', in Hallebeek, P.B., ed., *Symposium on Ethnographic and Water-logged Leather*, International Committee for Conservation, Amsterdam, 61–9.

Yeager, W.E., 2005, *A survey of stabilisation and storage techniques to protect iron encrustations and artefacts recovered from salt water marine archaeology sites* (http://ieeexplore.ieee.org – use 'advanced search' on webpage to find this paper).

Index

www.ingramcontent.com/pod-product-compliance
Lightning Source LLC
LaVergne TN
LVHW071217210926
848482LV00015B/831